LINGUA MEDICA

[An Indispensable resource for Students of Allied Health Professions]

By

FOLASAYO AYEGBAYO. DBA, Ph.D, FCIML

**YABA
LAGOS NIGERIA**

First Published in 2019
Copyright ©2019: Folasayo Ayegbayo
₊234 – 0802 – 757 – 3551

Contents

Preface

This book is especially designed for students of Allied Health Professions. It covers all the terminologies for all the body systems, both anatomically and patho-physiologically. It is an indispensable companion

Folasayo Ayegbayo.

CHAPTER ONE

INTRODUCTION TO MEDICAL TERMINOLOGY

LEARNING OBJECTIVES

After completing this chapter, students will be able to:

- *Apply the phonetic pronunciation guides that are used in frames.*

- *Recognize that medical terminology has both constructed and non-constructed terms.*

- *Identify each of the three word parts (word roots, prefixes, and suffixes) used to construct medical terms.*

- *Identify the function of a combining vowel that is added to a word root to form a combining form.*

- *Recognize that many medical terms are constructed from word parts and can be deconstructed into their word parts.*

⟹ **Factoid**

Most medical terms come from Latin and Greek origins.

- **Overview of Introduction to Medical Terminology**

Word Parts are the Key!- Introduction to word parts and how they create complex terms.

Word Roots - The word parts that usually, but not always, indicate the part of the body involved.

Combining Forms - Word roots plus a vowel (usually the letter **o**) added to the end. This form is used when connecting word roots or when the word root is joined to a suffix that begins with a consonant.

Suffixes - The word parts that usually, but not always, indicate the procedure, condition, disorder, or disease.

Prefixes - The word parts that usually, but not always, indicate location, time, number, or status.

Determining Meanings on the basis of Word Parts - Use knowledge of word parts to decipher medical terms.

Medical Dictionary Use - Guidelines to make the use of a medical dictionary easier.

Pronunciation - Learn the easy-to-use "sound-like" pronunciation system.

Spelling is always Important - Discover how one wrong letter can change the entire meaning of a term.

Using Abbreviations - Caution is important when using abbreviations.

Singular and Plural Endings - Usually singular and plural endings used in medical terms.

Basic Medical Terms - Terms used to describe disease condition.

Look-Alike Sound-Alike Terms and Word Parts - Clarification of confusing terms that look or soundalike.

Constructed Terms - These are terms which are made up of multiple word parts that are combined to form a new word.

Non-constructed Terms - are terms that are not formed from individual word parts. Non-constructed terms include eponyms (terms derived from the names of people).

Word Root: The part of a word that contains its primary meaning is the main body or core of the word. There is at least one per word.

arter	-	arteri artery	*norm*	-	a common state
arthr	-	joint	*oste*	-	bone
card, cardi-	-	heart	*path*	-	disease
gastr	-	stomach	*tens*	-	to stretch
hepat	-	liver	*ven*	-	vein
later	-	side			

Prefix: The word part that is placed before the root to modify its meaning.

ab-	-	away from	*hypo-*	-	below, under, deficient
ana-	-	up			
bi-	-	two	*intra-*	-	within
brady-	-	slow	*micro-*	-	small
endo-	-	within	*neo-*	-	new
epi-	-	upon, over, above, on top	*pre-*	-	to come before
			post-	-	to follow after
hyper-	-	above, beyond, excessive	*sub-*	-	under

Suffix: The word part that is attached to the end of the word to modify its meaning

-al	-	pertaining to	*-itis*	- inflammation
-ectomy	-	surgical excision		(swelling)
-emia	-	blood condition	*-logy* -	study of
-gnosis	-	knowledge	*-meter* -	measure
-gram	-	a record or image	*-pathy* -	disease
-iatry	-	treatment, specialty	*-plasty* -	surgical repair
-ic	-	pertaining to	*-scope* -	an instrument used for
-ist	-	one who specialized		viewing
-itis	-	inflammation	*-scopy* -	use of an instrument for
-ous	-	pertaining to		viewing

The **combining vowel** is used when a word root requires a connecting vowel to add a suffix or another word root when forming a term.

- In most cases, the combining vowel is the letter *o;* in some cases, it is the letter *i* or *e.*

- The word root plus combining vowel form is called a **combining form.**

Combining Vowel Rules

1. Are used between two word roots.

2. Are used between a word root and suffix.

3. Are used when the suffix or second word root begins with a consonant.

4. Are not used between a prefix and a word root.

5. Are used to ease pronunciation of a word.

Combining Forms occur when a word root is shown with the combining vowel attached, separated by a backslash.

arteri/o	-	artery	laryng/o	-	larynx, voicebox
appendic/o	-	appendix	leuk/o	-	white
arthr/o	-	joint	mamm/o	-	breast
bi/o	-	life	mast/o	-	breast
cardi/o	-	heart	ment/o	-	mind
cerebr/o	-	cerebrum	nat/o	-	birth
dermat/o	-	skin	neur/o	-	nerve
electr/o	-	electricity	oste/o	-	bone
encephal/o	-	brain	path/o	-	disease
gastr/o	-	stomach	proct/o	-	rectum
hem/o	-	blood	psych/o	-	mind
hepat/o	-	liver	rhin/o	-	nose
hyster/o	-	uterus	vas/o	-	vessel
			ven/o	-	vein

Recognize that many medical terms are constructed from word parts and can be deconstructed into their word parts.

Additional words not built from word parts:

abate	-	to lessen, decrease, or cease
abscess	-	a localized collection of pus, which can occur in any part of the body
acute	-	sudden, sharp, severe; a disease that has a sudden
onset, severe symptoms,	-	and a short course

afferent	-	carrying impulses toward a center
ambulatory	-	able to walk
antidote	-	a substance given to counteract poisons and their effects
apathy	-	a condition in which one lacks feelings and emotions and is indifferent
chronic	-	pertaining to time; a disease that continues over a long time, showing little change in symptoms or course
disease	-	lack of ease
efferent	-	carrying impulses away from a center
empathy	-	a state of projecting one's own personality into the personality of another to understand the feeling, emotions, and behavior of the person
febrile	-	pertaining to fever
gram	-	a unit of weight in the metric system; equal to a cubic centimeter or a milliliter of water
illness	-	a state of being sick
liter	-	a unit of volume in the metric system; equal to 33.8 fluid ounces or 1.057 quarts
malaise	-	a bad feeling; a condition of discomfort, uneasiness; often felt by a patient with a chronic disease
pallor	-	paleness; a lack of color
rapport	-	a relationship of understanding between two individuals, especially between the patient and the physician
syndrome	-	a combination of signs and symptoms occurring together that characterize a specific disease
thermometer	-	an instrument used to measure degree of heat
topography	-	a recording of a special place of the body
triage	-	the sorting and classifying of injuries to determine priority of need and treatment

CHAPTER ONE

INTRODUCTION TO MEDICAL TERMINOLOGY

Worksheet 1

Phonetic Spelling Challenge

Spell the medical term correctly in the space provided.

1. proh NUN see AYE shun _____

2. phoh NET ik   _____

3. kar dee ALL oh jee   _____

4. GAS troh heh PAT ik   _____

5. oss tee oh PATH ik   _____

6. pee dee ah TRI shun   _____

7. FRAK sher   _____

8. bak ter ee YOO ree ah   _____

9. men IN goh seel   _____

10. limm FOH mah   _____

Spelling Challenge

These terms are spelled incorrectly. Spell each term correctly in the space

provided.

1. Sufix   _____

2. Preffix   _____

3. Epillepsy   _____

4. Neonateologist   _____

5. Proctoescopey   _____

6. Glueteus maximuse   _____

7. Salbingo   _____

8. Electrocardeogram   _____

9. Salpingooforectomy _____

10. Prenadal   _____

True/False

Mark each statement as true (T) or false (F).

_____   1. The combining vowel is always used at the end of a word

root to form a   combining form.

_____   2. The term musculoskeletal means "pertaining to muscular

and skeletal."

_____   3. Not every medical term has all three word parts.

_____   4. Medical terminology is a functional language.

_____   5. The word root and combining vowel hyster/o mean "uterus."

_____   6. The combining form path/o means "nerve."

_____   7. The combining form leuk/o means "disease."

_____   8. The combining form neur/o means "birth."

_____   9. The combining form mamm/o means "breast."

_____   10. The combining form proct/o means "rectum."

Fill in the Blank

Fill in the blank with the correct medical term from this chapter.

11. _____ is a prefix that means "small."

12. _____ is a prefix that means "above" or "on top."

13. The prefix brady- means _____.

14. The prefix pre- means _____.

15. The prefix neo- means _____.

16. The suffix -emia means _____.

17. The suffix -scope means _____.

18. The suffix -ic means _____.

19. The prefix endo- means _____.

20. The suffix pathy- means _____.

Word Search

Fill in the blank with the correct medical term from this chapter.

1. A(n) _____ is a word part that is affixed to the beginning of a word. 2. A(n) _____ is a word part that provides the primary meaning of the term. 3. As a general rule, the _____ is used only to connect a word root with a suffix that begins with a consonant. 4. The term _____ means "the condition of having gallstones." 5. The term _____ means "inflammation of the stomach and small intestine." 6. The term that means "inflammation of the appendix" is _____. 7. The term that means "surgical repair of the heart" is _____. 8. The suffix -ectomy means

_____.9. _____ means "a tumor in the bile vessel or duct."

10. Many medical terms are _____, or made up of multiple word parts that are combined to form a new word.

11._____ are terms that are not formed from individual word parts. 12. _____ means "inflammation of the heart."13. Inflammation of the brain is known as _____.14. A(n) _____ is a physician who specializes in the diseases of the heart. 15.The word root and combining vowel rhin/o mean _____.

CHAPTER TWO

UNDERSTANDING SUFFIXES

LEARNING OBJECTIVES

After completing this chapter, students will be able to:

- Define and spell the suffixes significant in medical terminology.

- Identify suffixes in medical terms.

- Use suffixes to build medical terms that pertain to medical specialties, conditions, and diseases.

DEFINE AND SPELL THE SUFFIXES SIGNIFICANT IN MEDICAL TERMINOLOGY.

Suffix guidelines:

- If the **suffix** begins with a vowel, drop the combing vowel from the combing form and add the **suffix.**

- If the **suffix** begins with a consonant, keep the combining vowel and add the **suffix** to the combining form.

- Keep the combining vowel between two or more roots in a term.

⇨ **Factoid**

- A spelling error that changes just one or two letters in a term can completely change its meaning.

- Most medical terms are formed by assembling various word parts to construct a term.

Suffixes of the Human Body

SUFFIX	DEFINITION
-a	singular
-ac	pertaining to
-ad	toward
-ade	process
-al	pertaining to
-algesia	pain
-algia	condition of pain
-ar	pertaining to
-ary	pertaining to
-asthenia	weakness
-atresia	closure
-capnia	condition of carbon dioxide
-cele	protrusion, hernia, swelling
-centesis	surgical puncture
-clasia	break apart
-clasis	break apart
-cele	protrusion
-crit	to separate
-desis	surgical fixation, fusion
-drome	run or running
-dynia	pain
-ectasis	dilation
-ectomy	surgical removal
-emesis	vomiting
-emetic	vomiting
-genic	pertaining to formation
-gram	a record or image
-graphy	recording process

-hemia	blood (condition of)
-ia	condition of
-ial	pertaining to
-iasis	condition of
-iatry	treatment, specialty
-ic	pertaining to
-ist	one who practices
-ion	process
-is	pertaining to
-ism	condition of
-lepsy	seizure
-lexia	word, phrase
-lith	stone
-logist	one who studies
-logous	pertaining to study
-logy	study of
-lysis	loosen, dissolve
-lytic	loosen, dissolve
-malacia	softening
-megaly	large
-metry	measurement
-noia	mind
-oid	resembling
-oma	tumor
-opia	vision
-opsy	view
-optosis	condition of falling or drooping
-osis	condition of
-oxia	condition of oxygen
-penia	abnormal reduction in number, deficiency

-pepsia	digestion
-pexy	surgical fixation, suspension
-phagia	swallowing
-phasia	speaking
-phil and -philia	loving or affinity for
-phobia	fear
-phonia	condition of sound or voice
-phylaxis	protection
-physis	to grow or growth
-plasia	shape, formation
-plasty	surgical repair
-plegia	paralysis
-pnea	breathing
-poiesis	formation
-practic	practice
-ptosis	drop down
-ptysis	to spit out fluid
-rrhage	bleeding
-rrhagia	condition of profuse bleeding or hemorrhage
-rrhagic	pertaining to bleeding
-rrhaphy	suturing
-rrhea	excessive discharge
-rrhexis	rupture
-s	plural
-salpinx	trumpet; fallopian tube
-sclerosis	condition of hardening
-sis	state of
-spasm	sudden involuntary muscle contraction
-stasis	standing still, dripping

-therapy	treatment
-tocia	birth or labor
-tome	cutting instrument
-tomy	incision
-tripsy	surgical crushing
-troph and –trophy	development
-urea	urine
-uresis	urination
-uria	urine, urination
-y	process of

IDENTIFY SUFFIXES IN MEDICAL TERMS.

Other Terms with Suffixes

MEDICAL TERM	*DEFINITION*
abrasion	the process of scraping away from a surface, such as skin or teeth, by friction.
anesthetize	to induce a loss of feeling or sensation with the administration of an anesthetic.
arousal	pertaining to a state of alertness.
asymmetrical	unequal in size or shape. Without proportion of the body or parts of the body; different in placement or arrangement about an axis.
asystole	literally means "without contraction of the heart;" a life-threatening cardiac condition characterized by the absence of electrical and mechanical activity in the heart.
comatose	pertaining to a state of deep sleep (coma).
epithelium	the structure that covers the internal and external organs of the body and the lining of vessels, body cavities, glands, and organs.

exogenous	originating outside the body or an organ of the body, or produced from external causes, such as a disease caused by a bacterial or viral agent foreign to the body.
gynecoid	resembling a female.
hypertrophy	excessive nourishment; the increase in the size of an organ, structure, or the body, caused by an increase in the size of the cells rather than in the number of cells; overgrowth.
infection	process whereby a pathogenic microorganism invades the body, reproduces, multiples, and causes disease.
irregular	pertaining to not being regular.
palpate	to use the hands or fingers to examine by touch; to feel
steroid	resembling a solid substance; applies to any one of a large group of substances chemically related to sterols.
trauma	a physical injury or wound caused by external force, violence, or a toxic substance.

CHAPTER TWO

UNDERSTANDING SUFFIXES

Worksheet 1

Spelling Challenge

These terms are spelled incorrectly. Spell each term correctly in the space provided.

1. Pleuorocentesis _____

2. Lithotripsey _____

3. Angeography _____

4. Cystonscope _____

5. Chiropractice _____

6. Lapraroscopy _____

7. Angorrhaphy _____

8. Osteoclasious _____

9. Bronkospsm _____

10. Hydrphobia _____

11. Neuroarthropathey _____

12. Gastrostomey _____

13. Rhinorhagia _____

14. Menigeocele

_____

15. Arterioclerosis

_____

True/False

Mark each statement as true (T) or false (F).

_____ **1.** The suffix -*asthenia* means "weakness."

_____ **2.** The suffix -*pathy* is very common, and means "disease."

_____ **3.** The suffix -*pexy* means "surgical fixation or suspension."

_____ **4.** The suffix -*spasm* indicates "a sudden, involuntary muscle contraction."

_____ **5.** The suffix -*ab* means "toward."

Fill in the Blank

Fill in the blank with the correct medical term from this chapter.

11. A(n) _____ is a procedure that involves fusing together two or more vertebrae in the spine using either bone grafts or metal rods.

12. A(n) _____ is a procedure in which the gallbladder is excised or removed.

13. The term _____ means "the narrowing or

constriction of the trachea."

14. The term _____ means "vomiting of blood."

15. This term is defined as a difficulty in swallowing.
_____

16. In the term _____, the suffix -*phonia* means "sound" or "voice."

17. In the term _____, the suffix indicates the formation of blood cells.

Short Answer

Write the definition for each of the following terms.

21. Toxemia _____

22. Myasthenia _____

23. Rhinorrhagia _____

24. Calcipenia _____

25. Amniorrhexis _____

Word Search

Fill in the blank with the correct medical term from this chapter, then find those terms in the word search puzzle that follows.

1. A(n) _____ is the word part that is attached to the end of the word root.

2. The suffix -*al* means _____.

3. The suffix -*logy* means _____.

4. The suffix -*meter* means _____.

5. The suffix -*pathy* means _____.

6. The suffix -*emesis* means _____.

7. The suffix -*malacia* means _____.

8. The suffix -*phagia* means _____.

9. The suffix -*phasia* means _____.

10. The suffix -*plasia* means _____.

CHAPTER THREE

UNDERSTANDING PREFIXES

LEARNING OBJECTIVES

After completing this chapter, students will be able to:

- Define and spell the prefixes commonly used in medical terminology.

- Identify prefixes in medical terms.

- Use prefixes to build medical terms.

DEFINE AND SPELL THE PREFIXES COMMONLY USED IN MEDICAL TERMINOLOGY.

Prefix guidelines:

- The term **prefix** means to fix before or to fix to the beginning of a word.

- A **prefix** can be a syllable or a group of syllables.

- **Prefixes** are united with or placed at the beginning of words to alter or modify their meanings or to create entirely new words.

- Not all medical words have a **prefix**, but when they do, the **prefix** will alter or modify the meaning of the word.

Other Prefixes of the Human Body

PREFIX	DEFINITION
ad-	toward
an-	without
ana-	up, toward
ante-	before
anti-	against
ambi-	both
aut-	self
bi-	two
brady-	slow
circum-	around
con-	with or together
contra-	counter or against
di-	double or two
dia-	through
dys-	bad, abnormal, painful, or difficult
ec-, ecto-	outside, out
ep-	upon, on, over
epi-	upon
eso-	inward
eu-	normal, good
ex-, exo-	outside, away from
extra-	outside
hemi-	one-half
heter-	different
homo-	same
infer-	below
inter-	between
iso-	equal

macro-	large
mal-	bad
megalo-	large, great
meta-	after, change
micro-	small
mono-	one
multi-	many
neo-	new
nulli-	none
oligo-	little
pan-	all, entire
para-	alongside, abnormal
per-	through
peri-	around, about
poly-	many, excessive or over
primi-	first
pro-	forward, preceding
pseudo-	false
pyo-	pus
quad-	four
re-, retro-	back
semi-	half
sym-	together, joined
syn-	together, joined
tachy-	rapid, fast
tetra-	four
trans-	through, cross/across, or beyond
tri-	three
uni-	one

IDENTIFY PREFIXES IN MEDICAL TERMS.

Many medical words are formed by uniting various prefixes with a single suffix like -pnea. Example:

1. apnea
2. bradypnea
3. dyspnea
4. eupnea
5. hyperpnea
6. hypopnea
7. tachypnea

Other Terms with Prefixes

MEDICAL TERM	DEFINITION
afebrile	without fever
anicteric	without jaundice
arrest	to stop, inhibit, restrain
bifurcate	having two forks or two branches or two divisions; forked
binary	separated into two branches, or composed
concentration	in psychology, the process of being able to bring to the center one thought and to focus upon it, while at the same time excluding other thoughts of two elements.
enucleate	literally means "to remove the kernel of." It is used to describe the removal of the eyeball.
extraocular	outside the eye, as used in describing the extraocular eye muscles.
hyperactive	the nature or quality of excessive activity
hypoplasia	underdevelopment of a tissue, organ, or body.

multifocal	pertaining to or arising from many locations
occlusion	the process of closing or state of being closed of a passage.
parasternal	pertaining to beside the sternum (breastbone)
pericardial	pertaining to the pericardium (sac surrounding the heart).
polydactyly	pertaining to having more than the normal number of fingers and toes.
Premenstrual	pertaining to the time before the discharge of the menses.
react	literally means "to act again"; to respond to a stimuli; to participate in a chemical reaction.
regurgitation	the process of a backward flow of solids or foods from the stomach to the mouth, or the back flow of blood through a defective heart valve.
Subacute	literally means "below sharp"; it describes a state between acute and chronic, with some acute features.
superinfection	a new infection caused by a different organism from that which caused the initial infection
unconscious	not aware.

USE PREFIXES TO BUILD MEDICAL TERMS

Content Abstract

- To be able to identify the correct meaning of the prefix, you will need to analyze the definition of the medical word.

- Prefixes that carry meanings such as *away from, toward, before, above*, and *below* are often combined with roots and suffixes to describe a position or placement.

- Prefixes with meanings such as *both, ten, double, many, half,* and *none* are often combined with roots or suffixes to describe numbers or amounts.

CHAPTER THREE

UNDERSTANDING PREFIXES

Worksheet 1

Spelling Challenge

These terms are spelled incorrectly. Spell each term correctly in the space provided.

1. Intraoccular _____

2. Diplegea _____

3. Pollydipsia _____

4. Ambydeckstrus _____

5. Nulegravida _____

6. Monopleegia _____

7. Quadripleasia _____

8. Prenatul _____

9. Antennatal _____

10. Subcutaneus _____

True/False

Mark each statement as true (T) or false (F).

_____ **1.** The prefix *hyper-* means "excessive," or "beyond."

_____ **2.** The prefix *anti-* means "against" or "opposite."

_____ **3.** The prefix *di-* means "triple."

_____ **4.** The prefix *primi-* means "first."

_____ **5.** The prefix *uni-* means "one."

_____ **6.** The prefix *hemi-* means "two."

_____ **7.** The prefix *ambi-* means "self."

_____ **8.** The prefix *circum-* means "through."

_____ **9.** The prefix *pyo-* means "fire."

_____**10.** The prefix *ab-* means "away from."

Fill in the Blank

Fill in the blank with the correct medical term from this chapter.

11. The term *nulligravida* means _____ pregnancies.

12. A percutaneous drug is administered _____ the skin.

13. Because the prefix *inter-* means "between," the term _____ indicates a position between the vertebrae.

14. The prefixes *ec-* and *ecto-* mean _____.

15. The prefix *ad-* means "toward, near," or "increase," so the term *adduction* means _____ the midline of the body.

16. Paralysis of corresponding parts on both sides of the body, such as both arms or both legs, is called _____.

17. The prefixes *ex-* and *exo-* share the meaning "away from," as in the term *exotropia*, which is when the eye deviates _____ its normal position.

18. The prefixes *super-* and *supra-* share the meaning _____.

19. The prefix *pan-* means _____.

20. The prefix *eso-* means _____.

Short Answer: *Write the definition for each of the following terms.*

21. Dialysis _____

22. Dystrophic _____

23. Hyperinsulinism _____

24. Tachycardia _____

25. Bradycardia _____

CHAPTER FOUR

THE HUMAN BODY IN HEALTH AND DISEASE

LEARNING OBJECTIVES

After completing this chapter, students will be able to:

- Define and spell the word parts used to create terms for the human body.

- Identify the building blocks, organ systems, and cavities of the body.

- Identify the anatomical planes, regions, and directional terms used to describe areas of the body.

- Break down and define the important terms associated with the anatomy and physiology of the human body.

- Define the introductory terms associated with medical terminology.

- Identify the five major diagnostic imaging procedures.

⟹ Factoid

The word *etiology* is mainly used in medicine, where it is the science that deals with the causes or origin of disease, the factors that produce or predispose toward a certain disease or disorder.

DEFINE AND SPELL THE WORD PARTS USED TO CREATE TERMS FOR THE HUMAN BODY.

COMBINING FORM	DEFINITION
abdomin/o	abdomen
anter/o	front
brachi/o	arm
caud/o	tail
cephal/o	head
cervic/o	neck
cran/o, crani/o	skull
cyt/o	cell
dist/o	distant
dors/o	back
femor/o	thigh
gastr/o	stomach
glute/o	buttock
hom/o, home/o	same
ili/o hip,	groin
infer/o	below
inguin/o	groin
lumb/o	loin or lower back
medi/o	middle
organ/o	tool
pelv/o	bowl
physi/o	nature
poster/o	back
proxim/o	near
superi/o	above
thorac/o	chest, thorax
tom/o	to cut

umbilic/o	navel
ventr/o	belly

The basics:

- Medical terminology is the study of terms that are used in the art and science of medicine.
- The fundamental elements in medical terminology are the component parts used to build medical words.
- A root is a word or word element from which other words are formed. It is the foundation of the word and conveys the central meaning of the word.
- CF stands for combining form—a word root to which a vowel has been added to join the root to a second root or to a suffix.
- The component parts are P for prefix, R for root, CF for combining form, and S for suffix.
- These component parts are arranged according to the cell, tissue, organ, system, or element they describe.

IDENTIFY THE BUILDING BLOCKS, ORGAN SYSTEMS, AND CAVITIES OF THE BODY

CF/DEF	Medical Term	Definition
abdomin/o (*abdomen*)	abdominal cavity	the space inside the belly or abdomen; viscera; include the stomach, pancreas, spleen, liver, and most of the intestines
	abdominopelvic	the body cavity encompassing both

	cavity	the abdominal and pelvic cavities
anter/o (*front*)	anterior	pertaining to the front
brachi/o (*arm*)		pertaining to the arm
cardi/o (*heart*)	cardiovascular system	the body system that circulates blood throughout the body via the heart and blood vessels
caud/o (*tail*)	caudal	pertaining to the tail
cephal/o (*head*)		pertaining to the head
cervic/o (*neck*)		pertaining to the neck
chondr/o (*cartilage*)		part of the abdominal region
cran/o, crani/o (*skull*)	cranial	pertaining to the head
dist/o (*distant*)	distal	pertaining to away from a point of reference
dors/o (*back*)	dorsal	pertaining to the back
	dorsal cavity	containing the cranial and spinal cavities
gastr/o (*stomach*)		pertaining to the belly
femor/o (*hip*)		pertaining to the thigh

glute/o (*buttock*)		pertaining to the buttock
hom/o, home/o (*same*)	homeostasis	the process of maintaining internal stability
ili/o (*hip, groin*)		pertaining to the groin
infer/o (*below*)	inferior	pertaining to below a reference point
inguin/o (*groin*)		pertaining to the groin
lumb/o (*loin or lower back*)		pertaining to the lower back
medi/o (*middle*)	medial	toward the middle
	mediolateral	pertaining to the middle and to the side
pelv/o (*washbasin, pelvis*)	pelvic cavity	the space inside the pelvic area; viscera include urinary bladder, part of intestines, and internal reproductive organs
physi/o (*nature*)	physiologist	one who studies the nature of living things
	physiology	the study of the nature of living things
poster/o (*back*)	posterior	pertaining to the back

	posterolateral	pertaining to the back and to the side
	posteroanterior	pertaining to the back and front
proxim/o (*near*)	proximal	pertaining to near to a point of reference
spin/o (*spine* or *thorn*)	spinal cavity	the space inside the spinal cavity (also called the vertebral canal); viscera include the spinal cord
superi/o (*above*)	superior	pertaining to above a reference point
	superolateral	pertaining to above and to the side
tom/o (*to cut*)	anatomical position	the body position that is used as a reference for directional terms; it is an erect position with the arms at the side, palms of the hands facing forward, and legs together with feet pointing forward
	anatomy	the science of body structure
thorac/o (*chest, thorax*)	thoracic cavity	the space inside the chest or thorax; viscera include the heart, aorta, lungs, esophagus, bronchi
umbilic/o (*navel*)		pertaining to the navel

ventr/o (*belly*)	ventral	adjective meaning "pertaining to the belly"
	ventral cavity	contains the thoracic and abdominopelvic cavities

IDENTIFY THE ANATOMICAL PLANES AND REGIONS AND DIRECTIONAL TERMS USED TO DESCRIBE AREAS OF THE BODY.

Medical Regions

Medical Term	Definition
abdominal region	the belly region or abdomen, which contains several more specific regions, including the pigastric, hypogastric, hypochondriac, iliac, lumbar, and umbilical regions
brachial region	the region of the body pertaining to the arm; also called the arm region
cephalic region	the region of the body pertaining to the head; also called the head region
cervical region	the region of the body pertaining to the neck; also called the neck region
femoral region	the region of the body pertaining to the thigh; also called the thigh region

gastric region	part of the abdominal region specific to the area above the belly
gluteal region	the region of the body pertaining to the buttock; also called the buttock region
hypochondriac	part of the abdominal region specific to the area just below the cartilage of the ribs, which includes a right and left hypochondriac region on either side of the epigastric region; also called the infrachondrial region
hypogastric region	part of the abdominal region specific to the area below the belly
left iliac region	part of the abdominal region specific to the area located to the left of the hypogastric region
right iliac region	part of the abdominal region specific to the area located to the right of the hypogastric region
inguinal region	the region of the body pertaining to the groin; also called the groin region
left lumbar region	part of the abdominal *lower back* region specific to the area located to the left of the umbilical region
right lumbar region	abdominal region specific to the area located to the right of the umbilical region

thoracic region	the region of the body pertaining to the chest; also called the chest region or thorax
umbilical region	the area of the abdomen that contains the navel

Medical Directions

Medical Term	Definition
anatomical position	an erect posture with the arms at the side, palms of the hands facing forward, and legs together with feet pointing forward
superior	toward the head end or upper part of the body
inferior	away from the head end or toward the lower part of the body
anterior	toward the front or belly side
posterior	toward the back
medial	toward the midline, which is an imaginary vertical line down the middle of the body
lateral	toward the side
superficial	external, toward the body surface

deep	internal, inward from the surface of the body
proximal	toward the origin of attachment to the trunk
distal	away from the origin of attachment to the trunk
ventral	pertaining to the belly
dorsal	pertaining to the back
cephalic	pertaining to the head
caudal	pertaining to the tail

BREAK DOWN AND DEFINE THE IMPORTANT TERMS ASSOCIATED WITH THE ANATOMY AND PHYSIOLOGY OF THE HUMAN BODY.

Symptoms and Signs

Medical Term	Definition
Fever	the presence of an elevated body temperature
Pain	an unpleasant sensory and emotional experience that is associated with tissue damage
Sensation	a feeling or mental experience perceiving any stimuli
Sign	a finding that can be discovered by an objective examination
Symptom	experiences of the patient resulting from a disease

Diseases and Disorders

Medical Term	Definition
Acute	describes a disease of short duration, often with a sharp or severe effect
Chronic	describes a disease of long duration
Disease	the condition of instability that results when body functions fail to maintain homeostasis
Iatrogenic	adjective meaning a disease that is induced by medical treatment
Idiopathic	adjective meaning a disease of unknown cause
Infection	occurs when parasitic organisms within the body, such as bacteria, viruses, and fungi, attack body cells
Inflammation	a response to a trauma that is marked by the symptoms of redness, swelling, heat, and pain
Sequelae	conditions following and resulting from a disease
Trauma	a physical injury

BUILD MEDICAL TERMS FROM THE WORD PARTS ASSOCIATED

WITH THE ANATOMY AND PHYSIOLOGY OF THE HUMAN BODY.

Medical Term	Definition
Appendage	the head, arms, and legs attached to the trunk of the body
Atom	a nonliving particle that is capable of combining with other atoms to form more complex structures
Cell	the most basic living unit
Diaphragm	a sheet of muscle that separates the thoracic and abdominal cavities
Digestive system	the body system that converts food into a form the body can use for energy, growth, and repair; its organs include the tongue, pharynx, esophagus, stomach, liver, salivary glands, pancreas, small intestine, large intestine, and rectum
Endocrine system	the body system that regulates body function by secreting hormones; its organs include the pituitary gland, thyroid gland, parathyroid glands, adrenal glands, pancreas, and gonads
Female reproductive system	the female body system that enables reproduction by producing germ cells; its organs include the ovaries,

uterus, fallopian tubes, vagina, and vestibular glands

Integumentary system	the body system that provides a barrier to protect against fluid loss, physical damage, and invasion by microorganisms; its major organ is the skin
Lymphatic system	organs include the spleen, thymus, lymph nodes, and lymphatic vessels
Male reproductive system	the male body system that enables reproduction by producing germ cells; its organs include the testes, epididymus, vas deferens, prostate gland, seminal vesicles, bulbourethral glands, and penis
Molecule	a nonliving particle that is capable of combining with other molecules to form more complex structures
Muscular system	the body system that enables complex movement; its primary organs are the muscles
Nervous system	the body system that enables perception through the senses, integrates information to form thoughts and memories, and controls body movement and many internal functions; its organs include the brain, spinal cord, and nerves
Organ	a structure made of two or more different types of tissue that performs a general function in the body

Organ system	a combination of organs and associated structures that share a common goal; there are eleven organ systems in the body
Organism	the whole, complete human body that is capable of survival
Respiratory system	the body system responsible for bringing oxygen into the bloodstream; its organs include the nose, pharynx, larynx, trachea, bronchi, and lungs
Skeletal system	the body system that provides structure and support for other systems and aids in movement; its organs include bones and joints
Tissue	a group of similar cells that share a common goal or function
Trunk	the torso of the body
Urinary system	the body system that performs waste excretion; its organs include the kidneys, ureters, urinary bladder, and urethra
Viscera	the internal contents of body cavities, including organs, fluid, and connecting structures

Fever occurs when the body's internal "thermostat" raises the body temperature above its normal level. This thermostat is found in the part of the brain called the *hypothalamus.* The hypothalamus knows what temperature your body should be (usually around 98.6° Fahrenheit, or about 37° Celsius) and will send messages to your body to keep it at that temperature.

IDENTIFY THE FIVE MAJOR DIAGNOSTIC IMAGING PROCEDURES.

Treatments, Procedures, and Devices

CAT scan	computed axial tomography—a diagnostic scan similar to X-ray, in which data from beams of energized particles (or X-rays) are computer interpreted to produce a three-dimensional, cross-sectional "slice" or image of the body
Diagnosis	the determination of the nature of a disease
Endoscopy	a diagnostic procedure involving a visual examination using an endoscope, which includes a camera, fiber optics, and a long, flexible tube that can be inserted into the patient
Etiology	study of the causes of disease
Examination	an evaluation made for the purpose of diagnosis by

identifying physical evidence of disease, such as signs and symptoms

MRI   magnetic resonance imaging—a diagnostic scan that uses a powerful magnetic field generated within a chamber in which a patient lies; the field traces the hydrogen in the patient's body, the results of which are the clearest, most complete computer generated three-dimensional images of soft tissue that are currently possible

Pathologist   one who studies disease

Pathology   the study of disease

PET scan   positron emission tomography—a diagnostic scan that employs computers and radioactive substances to examine the metabolic activity of various parts of the body and create color-coded images

Prognosis   a forecast of the probable cause or outcome of a disease

Ultrasound   sonography—a diagnostic procedure in which harmless
imaging   sound waves are pulsated through body tissue; the pulse echoes are converted into images of internal body structures by computer

CHAPTER FOUR

THE HUMAN BODY IN HEALTH AND DISEASE

Worksheet 1

Phonetic Spelling Challenge

Spell the medical term correctly in the space provided.

1. ann AH toe mee _____

2. fiz ee OL oh jee _____

3. SAJ ih tal _____

4. HOE mee oh STAY siss _____

5. mee dee ah STY num _____

6. ab DOMM ih nahl _____

7. HIGH poh GASS trik _____

8. ap PEN dah jiz _____

9. DYE ah fram _____

10. dih ZEEZ _____

Spelling Challenge

These terms are spelled incorrectly. Spell each term correctly in the space provided.

1. Illiac _____

2. Peracardiol cavity _____

3. Proximil _____

4. Mediastynum _____

5. Organeles _____

6. Abdomminal _____

7. Homostasis _____

8. Coronall _____

9. Plural cavity _____

10. Anatonical planes _____

True/False

Mark each statement as true (T) or false (F).

_____ **1.** Anatomical position is an erect posture with the arms at the side, palms of the hands facing forward, and legs together with the feet pointing forward.

_____ **2.** The abdominal cavity contains the urinary bladder, internal reproductive organs, and parts of the small and large intestines.

_____ **3.** The term *pelvic* literally means "pertaining to a bowl."

_____ **4.** The process of reducing blood loss is called hemostasis.

_____ **5.** The simplest building block of the body is known as the atom.

_____ **6.** Each lung is surrounded by the pleura, a double-layered membrane.

_____ **7.** Two or more different tissues combine to form an organ system.

_____ **8.** The lymphatic system removes unwanted substances and recycles fluid to the blood.

_____ **9.** *Superior* means "pertaining to the back."

_____**10.** The two main cavities of the body are the dorsal cavity and the vertebral cavity.

Fill in the Blank

Fill in the blank with the correct medical term from this chapter.

11. Literally meaning "the process of cutting up," the study of body structure is called _____.

12. The general function of the _____ system is to exchange gases between the external environment and the blood.

13. A frontal or _____ plane is a vertical plane passing through the body from side to side, dividing the body into anterior and posterior portions.

14. Dorsal and _____ are interchangeable terms meaning "pertaining to the back."

15. In general, the term _____ refers to a state of the body in which homeostasis has faltered due to any cause.

16. A _____ _____ is a diagnostic procedure that combines multiple X-rays and computer enhancement to produce three-dimensional images of internal body structures.

Short Answer

Write the definition for each of the following terms.

21. Epigastric _____

22. Diaphragm _____

23. Sign _____

24. Chronic _____

25. Lateral _____

Word Search

Fill in the blank with the correct medical term from this chapter, then find those terms in the word search puzzle that follows.

1. There are four main categories of _____: epithelial, connective, muscle, and nervous.

2. The function of the _____ is to convert food

material into a molecular form that can be absorbed by the bloodstream and transported to body cells for nourishment.

3. The general function of the _____ is to control homeostasis by releasing hormones into the bloodstream.

4. In addition to lymph nodes, the _____ includes numerous other organs, many of which consist of clusters of white blood cells, such as monocytes and lymphocytes.

5. The process of maintaining internal stability is a central concept of human physiology and is called _____.

6. The _____ is the simplest organized substance known, although it too is composed of smaller particles.

7. The _____ means "on top of the belly."

8. When looking at the body as a whole, you will notice that its basic design consists of a central trunk or torso with attached _____, or limbs.

9. A(n) _____ is a horizontal plane dividing the body into superior and inferior portions.

10. The primary function of the cardiovascular system is to _____ vital substances throughout the body.

CHAPTER FIVE

THE INTEGUMENTARY SYSTEM

LEARNING OBJECTIVES

After completing this chapter, students will be able to:

- Define the word parts used to create medical terms of the integumentary system.

- Break down and define common medical terms used for symptoms, diseases, disorders, procedures, treatments, and devices associated with the integumentary system.

- Build medical terms from the word parts associated with the integumentary system.

- Pronounce and spell common medical terms associated with the integumentary system.

⇒ **Factoids**

- People of African or Asian descent are more likely to get keloids than are people with lighter skin. (In this respect, keloids are exactly the opposite of most skin cancers, which tend to occur in light-skinned people and spare people of color.)

- All decubitus ulcers have a course of injury similar to a burn wound. This can be from a mild redness of the skin and/or blistering, such as a first-degree burn, to a deep open wound with blackened tissue, as in a third-degree burn. This blackened tissue is called eschar. Seborrheic dermatitis is a common disease that affects 2% to 4% of the general population. However, up to 85% of HIV-infected people experience seborrheic dermatitis at some time after they acquire the infection. The cause of seborrheic dermatitis is unknown, but many investigators believe the yeast Pityrosporum ovale plays a role in the disease.

- Kaposi's sarcoma, or KS, is a type of cancer that men with AIDS might develop. It is rarely seen in women. A superficial infection of axillary and pubic hairs that results in adherent yellow, black, or red concretions surrounding the hair shaft is often caused by gram-positive corynebacteria. The usually asymptomatic infection occurs in both temperate and tropical climates and is not limited by race or sex.

PRONOUNCE AND SPELL COMMON MEDICAL TERMS ASSOCIATED WITH THE INTEGUMENTARY SYSTEM.

The purpose of the skin is to:

1. Provide a physical barrier that protects against loss of body fluids, damage due to physical injury or ultraviolet light, and invasion of microorganisms

2. Help regulate body temperature

3. House sensory receptors that provide information about the outside environment—temperature, touch, pain, and pressure

4. Secrete fluids

BUILD MEDICAL TERMS FROM THE WORD PARTS ASSOCIATED WITH THE INTEGUMENTARY SYSTEM.

Word Root/Combining Vowel	Definition
aden/o	*gland*
aut/o	*self*
cutane/o	*skin*
cyan/o	*blue*
derm/o, dermat/o	*skin*
follicul/o	*follicle*
kerat/o	*horny tissue*
onych/o	*nail*
seb/o	*sebum, oil*

Abbreviation	Definition
BCC	basal cell carcinoma
bx	biopsy
SLE	systemic lupus erythematosus
SqCCa	squamous cell carcinoma
TBSA	total body surface area

BREAK DOWN AND DEFINE COMMON MEDICAL TERMS USED FOR SYMPTOMS, DISEASES, DISORDERS, PROCEDURES, TREATMENTS, AND DEVICES ASSOCIATED WITH THE INTEGUMENTARY SYSTEM

Signs and Symptoms

Medical Term	Definition
Abrasion	a skin wound caused by scraping
Abscess	a collection of pus from a localized infection
Alopecia	a loss or lack of scalp hair; also called "baldness"
Cellulite	a local uneven surface of the skin caused by fat deposition, usually in the thighs and buttocks
Cicatrix	a scar
Comedo	an elevated lesion formed from the buildup of sebum and keratin; also called a pimple
Contusion	an injury to the skin, causing discoloration and swelling without breaking the skin surface; also called a bruise
Cyanosis	a blue tinge of color to an area of the skin.
Cyst	a closed sac or pouch containing fluid
Edema	swelling caused by accumulation of fluid

Erythema	a general term for redness of the skin
Fissure	a narrow break or slit in the skin
Furuncle	a localized skin infection originating from a hair follicle
Induration	the formation of a local hard area on the skin or elsewhere
Jaundice	an abnormal yellow coloring of the skin; also called *xanthoderma*
Keloid	an overgrowth of scar tissue
Laceration	a torn or jagged wound
Lesion	a change in tissue due to disease or injury
Macule	a discolored flat spot, such as a freckle
Nevus	a circumscribed pigmented area, a mole, or a birthmark; the plural form is *nevi*
Pallor	abnormal lack of skin color; paleness
Papule	a small, solid, circumscribed skin elevation
Petechia	pinpoint skin hemorrhage; the plural form is *petechiae*
Pruritis	a symptom of itching
Purpura	a purple-red discoloration resulting from hemorrhage into the skin
Pustule	a small, circumscribed skin elevation that contains pus
Ulcer	an eroded lesion of the skin or mucous membrane
Urticaria	skin eruption, usually caused by an allergic reaction to food, infection,

or injury; also called hives

Verruca   a small, circumscribed skin elevation caused by a virus; also called a

                wart

Vesicle   small elevation of the epidermis that contains fluid; also called a

                blister

Wheal   a temporary, itchy elevation of the skin, usually with a white center

                and red perimeter; also called a welt

Diseases and Disorders

Prefix	Definition	Combining Form	Definition	Suffix	Definition
ec-	*out*	aden/o	*gland*	-a	*singular*
		actin/o	*radiation*		
par-	*alongside*	albin/o	*white*	-ia	*condition of*
		carcin/o	*cancer*	-ic	*pertaining to*
		cellul/o	*small cell*	-ism	*condition*
		chym/o	*juice*	-it is	*inflammation*
		crypt/o	*hidden*	-malacia	*softening*
		derm/o, dermat/o	*skin*	-oma	*tumor*
		follicul/o	*small follicle*	-osis	*condition of*

hidr/o	*sweat*	-pathy	*disease*
kerat/o	*horny tissue*	-rrhea	*excessive discharge*
leuk/o	*white*		
melan/o	*black*		
myc/o	*fungus*		
onych/o	*nail*		
pedicul/o	*body louse*		
scler/o	*thick, hard*		
trich/o	*hair*		
xer/o	*dry*		

Medical Term	**Definition**
Acne	an inflammatory eruption of the skin caused by bacterial infection of sebaceous glands and ducts
Actinic keratosis	a precancerous skin condition caused by exposure to sunlight. It is marked by overgrowth of the outer epidermal layer.
Albinism	a genetic condition characterized by the lack of production of melanin; an individual with this condition is referred to as an albino
Alopecia	a loss or lack of scalp hair; also known as baldness

Basal cell carcinoma	a tumor arising from the epithelium of the epidermis; it can spread locally if not treated, but seldom metastasizes
Burn	an injury to the skin caused by excessive exposure to fire, electricity, chemicals, or sunlight.
Carbuncle	a skin infection composed of a cluster of boils caused by staphylococci bacteria
Cellulitis	inflammation of connective tissue (in the dermis) caused by infection
Decubitus ulcer	a skin sore caused by pressure or immobility while lying down; also called a bedsore
Dermatitis	inflammation of the skin
Ecchymosis	a purplish patch on the skin caused by leaking blood vessels
Eczema	an inflammatory skin disease characterized by redness, blisters, scaling, and sensations of itching and burning
Erythroderma	abnormal redness of the skin
Herpes	a skin eruption characterized by clusters of deep blisters that appear periodically; there are many variations, all of which are caused by members of the virus family herpesvirus
Hidradenitis	inflammation of a sweat gland
Impetigo	contagious skin infection characterized by blisters that later erupt to

form a yellowish crust

Kaposi's sarcoma	a form of skin cancer characterized by the formation of purple or brown patches on the feet that spread by way of lymphatics; interpreted as a sign of AIDS
Leukoderma	abnormally light-colored skin
Melanoma	a malignant skin tumor that arises from melanocytes
Onychocryptosis	an ingrown nail (abnormally buried in skin)
Onychomalacia	softening of the nails
Onychomycosis	fungal infection of the nails
Paronychia	infection around the nail
Pediculosis	infestation of the hair and skin with lice
Psoriasis	a chronic skin condition characterized by red lesions covered with silvery scales
Scabies	skin eruption caused by the female itch mite, which burrows into the skin to extract blood; this disorder causes mild dermatitis
Scleroderma	thickening of the skin caused by swelling and thickening of fibrous connective tissue
Seborrhea	sebaceous gland hyperactivity, resulting in excessive discharge of sebum
Squamous cell	a skin cancer arising from the epidermis, it usually appears as a

carcinoma	firm, red elevation with scales; it grows relatively slowly, but is capable of metastasis in its later stages
Systemic lupus erythematosus	a chronic inflammatory disease of connective tissue affecting the skin and many other organs; its early stages are characterized by red patches on the face and joint pain; commonly called lupus
Tinea	a fungal infection of the skin; also called ringworm
Trichomycosis	fungus on the hair surface
Xeroderma	abnormally dry skin

Treatments, Procedures, and Devices

Prefix	Combining Form	Definition	Suffix	Definition
(none)	abras/o	*to rub away*	-ectomy	*surgical removal*
	aut/o	*self*	-ion	*process*
	derm/o, dermat/o	*skin*	-plasty	*surgical repair*
	rhytid/o	*wrinkle*	-tome	*a cutting instrument*

Medical Term Definition

Biopsy	surgical removal of tissue for evaluation
Debridement	removal of diseased or dead tissue and foreign matter from a wound

Dermabrasion	removal of skin scars with abrasives, such as sandpaper
Dermatoautoplasty	surgical repair using the patient's skin for skin graft; also called *autograft*
Dermatoheteroplasty	surgical repair using a skin source other than that of the patient for a skin graft; also called *allograft*
Dermatome	an instrument used to cut skin
Dermatoplasty	surgical repair of the skin
Emollient	an agent that softens or smoothes the skin
Rhytidectomy	excision of wrinkles
Rhytidoplasty	surgical repair of wrinkles

CHAPTER FIVE

THE INTEGUMENTARY SYSTEM

Worksheet 1

Phonetic Spelling Challenge: Spell the medical term correctly in the space provided.

1. kar sih NOH mah   _____

2. SEB or EE ik   _____

3. peh dik yoo LOH sis   _____

4. RIT ih doh PLASS tee   _____

5. RET ih noydz   _____

6. MELL ah nin   _____

7. sell you LYE tiss   _____

8. ak TIN ik kair ah TOH siss   _____

9. EK zeh mah   _____

10. imp eh TYE goh   _____

Spelling Challenge: These terms are spelled incorrectly. Spell each term correctly in the space provided.

1. tina capitise   _____

2. shinngles   _____

3. ackne   _____

4. careatosis   _____

5. folicullitis   _____

6. systemic lupuss erythermatosus   _____

7. hideradenitiss   _____

8. empetigo   _____

9. autograf   _____

10. rhythidoplasty   _____

Abbreviation Matchup: Select and match the correct abbreviation to the definition.

_____ **1.** basal cell carcinoma

_____ **2.** squamous cell carcinoma

_____ **3.** total body surface area

_____ **4.** systemic lupus erythematosus

_____ **5.** biopsy

  a. bx

  b. SLE

  c. BCC

  d. SqCCa

  e. TBSA

True/False

Mark each statement as true (T) or false (F).

_____ 1. Alopecia can be a sign of an infection of the scalp, of high fevers, or of emotional stress.

_____ 2. As some people age, their skin becomes lighter in color due to reduced activity of the pigment-producing cells in the skin, the melanocytes.

_____ 3. The condition psoriasis is a skin eruption caused by the female itch mite, which borrows into the skin to extract blood.

_____ 4. Rhytidoplasty is the surgical repair of skin wrinkles.

_____ 5. Liposuction is the removal of subcutaneous fat.

_____ 6. Topical and oral antibiotics are used to manage infections, such as acne and carbuncles.

_____ 7. Onychomalacia is an infection around the nail.

_____ 8. An ulcer is an erosion through the skin or mucous membrane.

_____ 9. A keloid is the common result of an injury caused by a tear, or perhaps a cut, by a sharp object with an irregular surface.

_____ 10. A discolored flat spot on the skin surface, such as a freckle, is clinically called a macule.

Fill in the Blank

Fill in the blank with the correct medical term from this chapter.

11. _____ is the yellowing that results from an abnormal release of pigments by the liver.

12. _____ the outermost organ of the body.

13. A clinical term for "scar" is _____.

14. A localized elevation of the skin that is a sign of a local inflammation is a(n)

    _____.

15. The clinical term for "pimple" is _____.

16. Commonly known as a bruise, a(n) _____ is a discoloration and swelling of the skin that is symptomatic of an injury, such as a blow to the body.

17. A closed sac or pouch on the surface of the skin that is filled with fluid is called a(n)

    _____.

18. If an abscess is associated with a hair follicle, the local swelling on the skin is called

    a(n) _____.

19. A(n) _____ is a general term describing any small, solid elevation on the skin.

20. _____ is an abnormally pale color of the skin.

Short Answer

Write the definition for each of the following terms.

21. Ecchymosis _____

22. Onychomalacia _____

23. Debridement _____

24. Dermatome _____

25. Xeroderma _____

Word Search

Fill in the blank with the correct medical term from this chapter, then find those terms in the word search puzzle that follows.

1. The most life threatening skin cancer is _____.

2. The Latin word for a parasitic body louse is _____.

3. The medical field that specializes in the health and disease of the integumentary system is known as _____.

4. As the outermost organ of the body, the skin is more exposed to the extremes of the external environment than any other organ, subjecting it to temperature fluctuations, physical injury, and _____.

5. The clinical term for a narrow break or slit in the skin is

_____.

6. The Greek word for "blush" is _____.

7. Similar to a macule but darker in color, a(n) _____ is a pigmented spot that is commonly called a mole.

8. A(n) _____ is a general term describing any small, solid elevation on the skin.

9. A temporary, itchy elevation of the skin, often with a white center and red perimeter, is called a(n) _____.

10. In general, any form of _____ produces a sign of scaly skin.

11. A genetic condition characterized by the reduction of the pigment melanin in the skin is known as _____.

12. _____ is an inflammation of the connective tissue in the dermis.

13. In the condition _____, the individual suffers from excessive perspiration.

14. In the nail condition _____, a nail becomes buried in the skin due to abnormal growth.

15. The condition _____ is a fungal infection of hair.

CHAPTER SIX

The Skeletal and Muscular Systems

LEARNING OBJECTIVES

After completing this chapter, students will be able to:

- Define and spell the word parts used to create medical terms for the skeletal and muscular systems.

- Break down and define common medical terms used for symptoms, diseases, disorders, procedures, treatments, and devices associated with the skeletal and muscular systems.

- Build medical terms from word parts associated with the skeletal and muscular systems.

- Pronounce and spell common medical terms associated with the skeletal and muscular systems.

⟹ Factoids

- The Greek physician Hippocrates (460–357 BC) wrote texts that discussed the importance of chiropractic care. He wrote, "Get knowledge of the spine, for this is the requisite for many diseases."

- Babies are born with about 300 to 350 bones, but many of these fuse together between birth and maturity to produce an average adult total of 206.

DEFINE AND SPELL THE WORD PARTS USED TO CREATE MEDICAL TERMS FOR THE SKELETAL AND MUSCULAR SYSTEMS.

The skeletal system consists of bones and joints.

Osteology—the study of bones

Arthrology—the study of joints or articulations

Osteologist—a professional who studies or treats bones

Arthrologist—a professional who studies or treats joints

Skeleton—the combination of bones and joints that performs the following important functions:

1. Support—The strong, rigid skeleton forms a structural frame that offers other body structures a sturdy place for support.

2. Protection—Some bones physically surround internal body organs; the hardness of bones provides a partial shield from injury to organs.

3. Aid in movement—Bones provide a place of attachment for skeletal muscles, enabling coordinated movement to occur. The rigid nature of bones provides leverage for attached muscles, giving them something firm to pull against during contraction.

4. Blood cell formation—Blood cells are manufactured by a blood-forming connective tissue called red bone marrow that is located within bone tissue.

5. Storage—Bone tissue is the storehouse and main reserve for two important minerals,

calcium and phosphate, that are needed for muscle contractions, nerve cell function, and movement of material across cell membranes.

Bones come in a variety of sizes and shapes, but all have similar internal organization.

The muscular system consists of more than 500 muscles. Their primary function is to produce movement and generate heat, which is the byproduct of the energy used during contraction. Important parts of the muscular system include

- Tendon—attaches muscles to bone

- Muscle—consists of long, rodlike cells, known as skeletal muscle fibers, that are bundled together to form the meat of the muscle

- Fascia—tough fibrous connective tissue that surrounds individual muscle cells, groups of cells known as muscle bundles, and the whole muscle

Muscle contraction occurs when all muscle fibers in a muscle shorten at the same time, causing the muscle to shorten in its overall length. Because the fascia of a muscle is continuous with a tendon, contraction of the muscle causes the muscle to pull on the tendon, increasing its tension. As a result, the tendon pulls on the bone to which it is attached.

Combining Form	Definition
arthr/o, articul/o	*joint*
burs/o	*purse or sac, bursa*
carp/o	*wrist*

chondr/o	*gristle, cartilage*
condyl/o	*knuckle of a joint*
cost/o	*rib*
cran/o, crani/o	*skull, cranium*
fasci/o	*fascia*
femor/o	*thigh, femur*
fibr/o	*fiber*
fibul/o	*fibula*
ili/o	*flank, hip, groin, ilium of the pelvis*
ischi/o	*hip joint, ischium*
menisci/o	*meniscus*
muscul/o	*muscle*
myel/o	*bone marrow*
my/o, myos/o	*muscle*
oste/o	*bone*
pariet/o	*wall*
patell/o	*patella*
petr/o	*stone*
phalang/o	*phalanges*

phys/o	*growth*
pub/o	*pubis*
radi/o	*radius*
sacr/o	*sacrum*
skelet/o	*skeleton*
spondyl/o, vertebr/o	*vertebra*
stern/o	*chest, sternum*
synov/o, synovi/o	*synovial*
tars/o	*tarsal bone*
ten/o, tendon/o	*tendon*

BREAK DOWN AND DEFINE COMMON MEDICAL TERMS USED FOR SYMPTOMS, DISEASES, DISORDERS, PROCEDURES, TREATMENTS, AND DEVICES FOR THE SKELETAL AND MUSCULAR SYSTEMS.

Prefix	Definition	Combining Form	Definition	Suffix	Definition
a-	*without*	arthr/o	*joint*	-algia, -dynia/ para>	*condition of pain*

brady-	*slow*	kinesi/o	*motion*	-a	*singular*
dys-	*bad, abnormal, painful, or difficult*	my/o	*muscle*	-ia	*condition of*
hyper-	*excessive, abnormally high, above*	tax/o	*reaction to a stimulus, movement*	-y	*process of*
		ten/o	*tendon*		
		troph/o	*development*		

Signs and Symptoms

Medical Term	Definition
Arthralgia	pain in a joint
Ataxia	an inability to coordinate muscles while executing a voluntary movement
Atrophy	lacking development, or wasting
Bradykinesia	abnormally slow movement
Decalcification	abnormal reduction of calcium in bone

Dyskinesia	difficulty in movement
Dystrophy	deformities arising during development
Hypertrophy	excessive development
Myalgia	muscle tenderness or pain
Tenodynia	pain in a tendon

Diseases and Disorders

Prefix	Definition	Combining Form	Definition	Suffix	Definition
a-	*without*				
epi-	*upon, above, over, or on top*	ankyl/o	*crooked*	-asthenia	*weakness*
para-	*beside, departure from normal*	arthr/o	*joint*	-cele	*protrusion*
poly-	*many*	burs/o	*purse or sac, bursa*	-genesis	*origin, cause*
quadri-	*four*	carcin/o	*cancer*	-itis	*inflammation*

carp/o	*wrist*	-malacia	*softening*
chondr/o	*cartilage*	-oma	*tumor*
condyl/o	*knuckle of a joint*	-osis	*condition of*
fibr/o	*fiber*	-plasia	*formation or growth*
kyph/o	*hump*	-plegia	*paralysis*
leuk/o	*white*		
lith/o	*stone*	-ptosis	*drooping, falling down*
lord/o	*bent forward*		
menisc/o, menisci/o	*meniscus*		
myel/o	*bone marrow*		
myos/o	*muscle*		
ost/o, oste/o	*bone*		
sarc/o	*flesh or meat*		
scoli/o	*curved*		
spondyl/o	*vertebra*		

synov/o, synovi/o   *synovial*

ten/o, tend/o   *tendon*

Medical Term	Definition
Achondroplasia	abnormal, slow growth of long bones, resulting in unusually short, stocky limbs
Ankylosis	an abnormal condition of joint stiffness
Arthritis	inflammation and degeneration of a joint
Arthrochondritis	inflammation of cartilages within joints
Bunion	abnormal enlargement of the joint at the base of the big toe
Bursitis	inflammation of a bursa
Bursolith	a calcium deposit within a bursa
Carpal tunnel syndrome	a repetitive stress injury in which the nerves of the wrist generate pain impulses due to inflammation of synovial sheaths
Carpoptosis	drooping of the wrist; also called wrist drop
Cramps	prolonged, involuntary muscular contractions
Duchenne's muscular dystrophy	a congenital condition resulting in progressive muscular weakness and deterioration
Epicondylitis	inflammation of the cartilages of the elbow

Fibromyalgia	a disease of unknown origin that produces widespread pain of musculoskeletal structures, other than joints, of the limbs, face, and trunk
Gout	abnormal deposition of uric acid crystals in the joints, causing localized pain; also called *gouty arthritis*
Herniated disk	a rupture of an intervertebral disk, resulting in the protrusion of tissue against spinal nerves, which generates pain
Kyphosis	a deformity of the spine characterized by the presence of a hump; also called *hunchback*
Lordosis	a deformity of the spine characterized by an anterior curve of the lumbar area
Marfan's syndrome	an inherited condition resulting in excessive cartilage formation in the epiphyseal plates, forming long arms and legs
Meniscitis	inflammation of a meniscus
Myasthenia gravis	weakness in the muscles
Myeloma	a malignant tumor of bone marrow
Myoclonus	a spasm or twitching of a muscle or muscle group
Myocele	protrusion of a muscle through its fascia
Myositis	inflammation of muscle tissue
Osteitis	inflammation of a bone

Osteitis deformans	viral infection of bone that causes deformities of the skeleton resulting from the acceleration of bone loss; also called *Paget's disease*
Osteoarthritis	a form of arthritis characterized by an age-related deterioration of joints that is accompanied by erosion of cartilage and painful inflammation
Osteocarcinoma	cancer of bone
Osteochondritis	inflammation of bone and associated cartilage
Osteogenesis imperfecta	an inherited condition resulting in impaired growth and fragile bones, leading to progressive skeleton deformation and frequent fractures
Osteomalacia	a gradual and painful softening of bones
Osteomyelitis	a painful bone infection caused by bacteria, characterized by inflammation of the red bone marrow
Osteonecrosis	death of bone tissue
Osteoporosis	abnormal loss of bone density
Osteosarcoma	cancer of bone
Paraplegia	paralysis of both legs and the lower part of the body
Polymyositis	inflammation of many muscles simultaneously
Quadriplegia	paralysis of all four appendages

Rheumatoid arthritis	a form of arthritis characterized by progressive, gradual joint deterioration that is caused by an autoimmune response
Rickets	the bones become softened due to excessive removal of calcium for other body functions.
Scoliosis	abnormal lateral curvature of the spine
Spinal cord injury	trauma to the spinal cord, often resulting in paralysis
Spondylarthritis	inflammation of the intervertebral joints
Sprain	an injury resulting from stretching a ligament beyond its normal range, tearing its collagen fibers
Strain	an injury resulting from stretching a muscle beyond its normal range, tearing muscle tissue
Tendonitis	inflammation of a tendon
Tenosynovitis	inflammation of a tendon and the synovial membrane that surrounds it

Fractures

Medical Term	Definition
Closed or simple fractures	fractures that are completely internal to the body
Open or compound fractures	fractures that project through the skin, outside the body
Colles fracture	a break in the distal part of the radius

Comminuted fracture	a break resulting in fragmentation of the bone
Compression fracture	a crushed break, often due to weight or pressure applied to a bone during a fall
Displaced fracture	a break causing an abnormal alignment of bone pieces
Epiphyseal fracture	a break at the location of the growth plate, which can affect growth of the bone
Greenstick fracture	a slight break in a bone that appears as a slight fissure in an X-ray
Nondisplaced fracture	a break in which the broken bones retain their alignment
Pott's fracture	a break at the ankle that affects both bones of the leg
Spiral fracture	a spiral-shaped break often caused by twisting stresses along a long bone

Treatments, Procedures, and Devices

Combining Form	Definition	Suffix	Definition
arthr/o	*joint*	-centesis	*surgical puncture*
burs/o	*purse or sac, bursa*	-clasia, -clasis	*break apart*
chondr/o	*cartilage*	-desis	*surgical fixation or fusion*
cost/o	*rib*	-ectomy	*surgical removal or*

			excision
crani/o	*skull, cranium*	-gram	*a record or image*
electr/o	*electricity*	-graphy	*recording process*
fasci/o	*fascia*	-iatry	*treatment or speciality*
lamin/o	*thin, lamina*	-ist	*one who specializes*
orth/o	*straight*	-lysis	*loosen or dissolve*
ost/o, oste/o	*bone*	-pathy	*disease*
spondyl/o	*vertebra*	-plasty	*surgical repair*
syn/o	*connect*	-rrhaphy	*suturing*
ten/o, tend/o	*tendon*	-scope	*instrument used for viewing*
vertebr/o	*vertebra*	-scopy	*process of viewing*
		-tomy	*incision or to cut*

Medical Term	Definition
Arthrocentesis	surgical puncture of a joint to aspirate fluid from a synovial cavity
Arthroclasia	the breaking of an abnormally stiff joint during surgery to increase range of motion
Arthrodesis	surgical fixation of a joint

Arthrogram	X-ray film of a joint after injection of air, contrast media, or both
Arthroplasty	surgical repair of a joint or total joint replacement
Arthroscopic surgery	surgery of a joint using the arthroscope
Arthroscopy	endoscopic visual examination of a joint cavity using a fiber-optic instrument, the arthroscope
Arthrotomy	surgical incision into a joint
Bursectomy	surgical excision of a bursa
Chiropractic	a field of therapy that mainly involves manipulation of the vertebral column
Chiropractor	a specialist in chiropractic
Chondrectomy	surgical excision of joint cartilage
Chondroplasty	surgical repair of joint cartilage
Costectomy	surgical excision of a rib
Cranioplasty	surgical repair of a defect in the cranium
Craniotomy	surgical entry into the cranium
Diskectomy	surgical excision of an intervertebral disk
Electromyography	a diagnostic procedure that records the strength of muscle contractions
Fasciotomy	surgical incision into fascia

Laminectomy	excision of part of a vertebra known as the lamina, often as part of a treatment for a herniated disk
Meniscectomy	surgical removal of a meniscus
Myoplasty	surgical repair of muscle tissue
Myorrhaphy	closing a muscle with sutures
Orthotics	construction and fitting of orthopedic appliances to assist in patient movement
Orthotist	a physician specializing in orthotics
Ostectomy	excision of bone
Osteoclasis	the surgical breaking of a bone to correct a deficiency
Osteopath	a physician trained in osteopathy; also called an *osteopathic surgeon*
Osteopathy	a medical field that focuses on the relationship between the musculoskeletal system and overall health, with an emphasis on preventative medicine
Osteoplasty	surgical repair of bone
Podiatrist	a physician trained in podiatry; also called a *chiropodist*
Podiatry	the medical field specializing in treating the foot
Prosthesis	an artificial substitute for a missing body part, such as a leg or hand
Tenomyoplasty	surgical repair of a muscle and its tendon

Tenorrhaphy	closing a tendon with sutures
Tenotomy	incision into a tendon
Vertebroplasty	a surgical procedure that repairs damaged or diseased vertebrae

Abbreviation	**Definition**
ACL	anterior cruciate ligament; a ligament that stabilizes the knee joint
CTS	carpal tunnel syndrome
DJD	degenerative joint disease
DMD	Duchenne's muscular dystrophy
DO	physician specializing in osteopathy
EMG	electromyography
HNP	herniated nucleus pulposus; a herniated intervertebral disk
MG	myasthenia gravis
NSAIDs	nonsteroidal anti-inflammatory drugs
OA	osteoarthritis
ortho	orthopedics
RA	rheumatoid arthritis
ROM	range of motion

SCI	spinal cord injury
THR	total hip replacement
TKA	total knee arthroplasty
TKR	total knee replacement
TMJ	temporomandibular joint

Vertebrae

C1 through C7	the seven cervical vertebrae
T1 through T12	the twelve thoracic vertebrae
L1 through L5	the five lumbar vertebrae

CHAPTER SIX

THE SKELETAL AND MUSCULAR SYSTEMS

Worksheet 1

Phonetic Spelling Challenge: Spell the medical term correctly in the space provided.

1. ah TAK see ah _____

2. DISS troh fee _____

3. TEN oh DINN ee ah _____

4. my AL jee ah _____

5. high PER troh fee _____

6. diss kih NEE see ah _____

7. ahr THRAL jee ah _____

8. BRAD ee kih NEE see ah _____

9. AT roh fee _____

10. DEE kal sih fih KAY shun _____

Spelling Challenge: These terms are spelled incorrectly. Spell each term correctly in the space provided.

1. spondilarthritis _____

2. tendonightis _____

3. arthrouplasty _____

4. electromiography

_____

5. kraniotomy

_____

6. carpull tunnel syndrome

_____

7. epicondilitis

_____

8. menisitis

_____

9. exsision

_____

10. tempoormandibular

_____

11. asspiration

_____

12. Archilly's tendon

_____

13. musculorskeletal

_____

14. scoleosis

_____

15. kiropractic

_____

Abbreviation Matchup: Select and match the correct abbreviation to the definition.

_____ **1.** Total knee replacement

_____ **2.** Myasthenia gravis

_____ **3.** Degenerative joint disease

_____ **4.** Range of motion

_____ **5.** Spinal cord injury

_____ **6.** Rheumatoid arthritis

_____ **7.** Total hip replacement

    a. ROM

    b. RA

    c. MG

    d. DJD

    e. TKR

    f. THR

    g. SCI

True/False

Mark each statement as true (T) or false (F).

_____ **1.** There are a total of 209 bones in the human body.

_____ **2.** Another term for a ruptured disk is a herniated disk.

_____ **3.** Inflammation of bone tissue is called osteitis.

_____ **4.** The term *myasthenia* means weakness in the muscles.

_____ **5.** A congenital disease called Marfan syndrome results in excessive cartilage formation at the growth plates, or epiphyseal plates, forming abnormally long limbs and a tall, thin body form.

_____ **6.** Kyphosis is an exaggerated anterior spinal curve in the lumbar area, causing the patient to bend forward.

_____ **7.** Scoliosis is a lateral curvature of the spine.

_____ **8.** Inflammation of the white bone marrow is a painful disease known as osteomyelitis.

_____ **9.** A sprain is a tear of collagen fibers within a ligament.

_____**10.** A sprain is a stretching of a muscle beyond its normal range.

Fill in the Blank

Fill in the blank with the correct medical term from this chapter.

11. _____ is the inflammation of a tendon, a common sports injury.

12. _____ is a form of tendonitis that involves inflammation of the synovial membrane.

13. A break or fracture at the ankle that affects both bones of the leg is called a _____.

14. When a break causes an abnormal alignment of bone pieces, it is called _____.

15. A crushed break is called a(n) _____.

16. The medical specialty that focuses on foot health is _____.

17. A medical field that emphasizes the relationship between the musculoskeletal system and overall health, with an emphasis on preventative herbal medicine, is _____.

18. A surgical procedure that repairs damaged or diseased vertebrae is called a(n)

_____.

19. The term that means withdrawal by suction is _____.

20. The correct medical term for "humpback" is _____.

Short Answer

Write the definition for each of the following terms.

21. Hypertrophy   _____

22. Dystrophy   _____

23. NSAIDs   _____

24. Ostectomy   _____

25. Myoplasty   _____

CHAPTER SEVEN

BLOOD AND THE LYMPHATIC SYSTEM

LEARNING OBJECTIVES

After completing this chapter, students will be able to:

- Define and spell the word parts used to create terms for the blood and the lymphatic system.

- Break down and define common medical terms used for symptoms, diseases, disorders, procedures, and treatments associated with the blood and the lymphatic system.

- Build medical terms from the word parts associated with the blood and the lymphatic system.

- Pronounce and spell common medical terms associated with the blood and the lymphatic system.

INSTRUCTIONAL GOAL: DEFINE AND SPELL THE WORD PARTS USED TO CREATE MEDICAL TERMS FOR THE BLOOD AND LYMPHATIC SYSTEM.

Content Abstract

1. Blood—vital body fluid that transports substances necessary for survival by way of the cardiovascular system. It consists of

   - Plasma—the yellowish fluid part of the blood

- Fibrinogen—protein of blood plasma that begins the blood-clotting process

- Serum—blood plasma that is lacking fibrinogen

- Formed elements—substances suspended in the fluid. The three types are

  - Red blood cells (RBCs), or erythrocytes—the most abundant cells that carry *hemoglobin,* a specialized protein that contains iron molecules. This enables the protein to bind to oxygen and carbon dioxide. They are produced by *stem cells* in the red bone marrow during *hematopoiesis.*

  - Platelets or thrombocytes—the second most abundant type of blood cell, platelets are actually fragments from huge cells that break apart during development in the bone marrow. They perform the role of preventing fluid loss by releasing proteins in a process known as *coagulation,* which results in the formation of *blood clots.*

  - White blood cells (WBCs), or leukocytes—the smallest group of cells in a normal blood sample, they perform an important role in protecting your body from infectious microorganisms and other foreign, unwanted materials. There are several types depending on their histological features:

    - Eosinophils—granulocytes, because they contain tiny, pebblelike objects in their cytoplasm. *(Their granules stain red.)* They actively attack and "eat" bacteria and unwanted cells in a process known as *phagocytosis.*

    - Basophils—granulocytes, because they contain tiny, pebblelike objects in their cytoplasm. (*Their granules stain blue.*) They release substances that trigger an allergic reaction.

- Neutrophils—granulocytes, because they contain tiny, pebblelike objects in their cytoplasm. (*Their granules stain pink in a neutral stain.*) Neutrophils, the most abundant, actively attack and "eat" bacteria and unwanted cells in a process known as *phagocytosis.*

- Monocytes—large *agranulocytes* that are aggressive "eaters" of bacteria and other unwanted cells, especially once they transform into *macrophages.*

- Lymphocytes—smaller agranulocytes that provide you with the most powerful immune reaction of the body. They include

  - T cells—capable of destroying unwanted substances by a variety of means and are important in activating the B cells.

  - B cells—produce the most effective weapon in the fight against infection—tiny molecules called *antibodies.* Antibodies attach to unwanted substances called *antigens,* rendering them ineffective.

2. The lymphatic system—closely associated with the blood and its circulation, the system also includes components that play a key role in protecting the body against infection. It consists of vessels and a yellowish liquid known as *lymph,* which flows in one direction toward the heart. The main functions of the lymphatic system are

- Recycling fluid back to the bloodstream

- Fighting infection with the white blood cells it contains

The organs include

- Microscopic lymphatic capillaries

- Lymphatic vessels—they deliver lymph into larger channels

- Lymphatic trunks

- Lymph nodes—pea-sized organs that lymph enters from the lymphatic vessels

- Spleen—located in the abdominal cavity

- Thymus gland—located in the chest

- Tonsils—three pairs in the throat

- Lymphatic nodes—embedded in the wall of the large intestines

The immune response is a mechanism in the body employed to battle *infections,* which are immunological diseases resulting from *pathogens (disease-causing agents that include viruses, bacteria, fungi, protozoans, and wormlike organisms).* Pathogens can cause harm in one of two ways:

- By destroying cells

- By releasing *toxins,* poisonous substances that interfere with cell function

The immune response is a series of reactions against infection that are orchestrated by white blood cells:

- Unwanted pathogens are attacked by white blood cells that are phagocytic. The process occurs mainly within the lymphatic organs. It often results in *inflammation,* which produces redness, swelling, heat, and pain at the infectious site.

If the infection is aggressive, phagocytosis cannot control the invaders alone. In this case, lymphocytes are brought to battle by chemical signals. The phagocytes produce chemical signals that result in rapid growth of lymphocyte populations. Two different mechanisms of lymphocyte activation take place during an infection. They are

- Innate immune response—T cells become activated, resulting in a rapid growth of the T cell population within lymphatic tissue. This further results in the formation of *specialized T cells* that can destroy the pathogens effectively. One population group is the:

  - Helper T cells—these are required for the activation of *B cells* during the second mechanism.

- Acquired immune response—during this reaction, B cells are transformed into *plasma cells,* which secrete enormous numbers of antibodies. Within days, the blood and lymph fill with these molecules, which render the antigen ineffective and halt the infection.

- Memory cells—following infection, certain lymphocytes remain in the bloodstream and lymph and "remember" the pathogen's molecular signal. This provides you with immunity.

Combining Forms

Combining Form	Definition
aden/o	*gland*
bacteri/o	*bacteria*
blast/o	*germ or bud, developing cell*
erythr/o	*red*
hem/o, hemat/o	*blood*
immun/o	*exempt or immunity*
leuk/o	*white*
lymph/o	*clear water or fluid*
path/o	*disease*
splen/o	*spleen*
thromb/o	*clot*
thym/o	*wartlike, thymus gland*
tox/o	*poison*

Factoid

Plasma is a straw-colored, clear liquid that is 90% water. Besides water, plasma contains dissolved salts and minerals like calcium, sodium, magnesium, and potassium. Microbe-fighting antibodies travel to the battlefields of disease by hitching a ride in the plasma.

INSTRUCTIONAL GOAL: BREAK DOWN AND DEFINE COMMON MEDICAL TERMS USED FOR SYMPTOMS, DISEASES, DISORDERS, PROCEDURES, TREATMENTS, AND DEVICES FOR THE BLOOD AND LYMPHATIC SYSTEM.

Signs and Symptoms

Prefix	Definition	Combining Form	Definition	Suffix	Definition
an-	*without or absence of*	bacteri/o	*bacteria*	-emia	*condition of blood*
iso-	*equal*	cyt/o	*cell*	-ia	*condition of*
macro-	*large*	erythr/o	*red*	-lysis	*loosen or dissolve*
poly-	*many*	hem/o	*blood*	-megaly	*abnormally large*
		leuk/o	*white*	-osis	*condition of*
		poikil/o	*irregular*	-penia	*abnormal reduction in number or deficiency*
		splen/o	*spleen*	-rrhage	*profuse bleeding, hemorrhage*
		thromb/o	*clot*		
		tox/o	*poison*		

Medical Term	Definition
Anisocytosis	presence of red blood cells of unequal size
Bacteremia	presence of bacteria in the bloodstream
Erythropenia	abnormally reduced number of red blood cells
Hemolysis	rupture of the red blood cell membrane
Hemorrhage	loss of blood from circulation
Leukopenia	abnormally reduced number of white blood cells
Macrocytosis	abnormally large red blood cells
Poikilocytosis	large, irregularly shaped red blood cells
Polycythemia	abnormal increase in the number of erythrocytes in the blood
Splenomegaly	abnormal enlargement of the spleen
Thrombopenia	abnormally reduced number of platelets
Toxemia	presence of toxins in the bloodstream

Diseases and Disorders

Prefix	Definition	Combining Form	Definition	Suffix	Definition
ana-	*up or toward*	aden/o	*gland*	-emia	*condition of blood*

an-	without or absence of	aut/o	self	-genic	pertaining to producing
mono-	one	fung/o	fungus		
		hem/o, hemat/o	blood	-ial	pertaining to
		globin/o	protein	-ic	pertaining to
		iatr/o	physician	-ism	condition
		idi/o	individual	-itis	inflammation
		immun/o	exempt or immunity	-oma	tumor
		leuk/o	white	-osis	condition
		lymph/o	clear water or fluid	-pathy	disease
		necr/o	death	-philia	loving or affinity
		nosocom/o	hospital		
		nucle/o	kernel or nucleus	-rrhagic	pertaining to profuse bleeding
		path/o	disease		
		sept/o	putrefying; wall or		

	partition
staphylococc/o	*Staphylococcus*
streptococc/o	*Streptococcus*
thym/o	*wartlike, thymus gland*

Medical Term	**Definition**
AIDS	the acronym for *acquired immune deficiency syndrome,* AIDS is caused by the *human immunodeficiency virus (HIV),* which disables the immune response by destroying mainly helper T cells (needed for activation of B cells); the loss of immune function allows opportunistic infections to proliferate and eventually cause death
Allergy	a response to an allergen, which is an antigen that produces a hypersensitivity reaction that includes immediate inflammation but does not elicit other immune responses. Allergies are of many types, the most common of which are *allergic rhinitis (hay fever),* which affects mucous membranes of the nasal cavity and throat, and *allergic dermatitis,* which affects the skin where it has made contact with the allergen.

Anaphylaxis	an immediate reaction to an antigen that includes rapid inflammation and systemwide smooth muscle contraction
Anemia	a reduced ability of red blood cells to deliver oxygen to tissue; common forms of anemia include *aplastic anemia, iron deficiency anemia, sickle cell anemia,* and *pernicious anemia*
Anthrax	a bacterial disease that has been threatened to be used in bioterrorism
Aplastic anemia	anemia characterized by the failure of red bone marrow to produce red blood cells
Autoimmune disease	any of several diseases that are caused by a person's own immune response attacking otherwise healthy tissues, including *rheumatoid arthritis, systemic lupus erythematosus,* and *multiple sclerosis*
Botulism	a form of poisoning caused by the ingestion of food contaminated with toxin produced by the bacterium *Clostridium botulinum*
Communicable disease	a disease that is capable of transmission from one person to another
Diphtheria	a disease caused by a bacterium and its toxin, resulting in inflammation of mucous membranes primarily in the mouth and throat
Dyscrasia	a general term for an abnormal condition of the blood
Edema	leakage of fluid from the bloodstream into the interstitial space between body cells causes swelling

Fungemia	a fungal infection distributed by way of the bloodstream
Gas gangrene	the infection of a wound, caused by various anaerobic bacteria, that produces a fermentation gas, necrosis, and septicemia
Hematoma	mass of blood outside of blood vessels and confined within an organ or space within the body, usually in clotted form
Hemoglobinopathy	general term for a disease that affects the hemoglobin within red blood cells
Hemophilia	an inherited bleeding disorder that results from defective clotting proteins involved in blood coagulation
Hemorrhagic fever	an infectous disease that causes internal bleeding or internal hemorrhage and high fevers
Hodgkin's disease	cancer of lymphatic tissue, characterized by the progressive enlargement of lymph nodes, fatigue, and deficiency of the immune response
Iatrogenic disease	a condition that is caused by a medical treatment
Idiopathic disease	a disease that develops without a known or apparent cause
Immunodeficiency	a condition resulting from a defective immune response
Immunosuppression	reduction of an immune response caused by disease or, in the case of organ transplant, by the use of chemicals, pharmacologic, physical, or immunologic agents

Incompatibility	the combination of two blood types that result in the destruction of red blood cells.
Infection	a multiplication of disease-causing microorganisms
Inflammation	a swelling of body tissue caused by movement of plasma into the extracellular space to produce *edema,* or fluid accumulation in tissue; symptoms include swelling, redness, heat, and pain
Influenza	a viral disease characterized by a temporary inflammation of mucous membranes and fever. Commonly called "the flu," it is highly contagious, and the virus is capable of mutating to escape detection by B and T memory cells
Iron deficiency anemia	anemia that is caused by a lack of iron, which results in smaller red blood cells containing deficient levels of hemoglobin
Leukemia	cancer of the red bone marrow, which is the blood-forming tissue
Lymphadenitis	inflammation of the lymph nodes
Lymphoma	a tumor originating in lymphatic tissue
Malaria	a disease caused by a parasitic protozoan carried by *anopheles* mosquitoes that infects red blood cells; it is characterized by periodic fever and fatigue
Mononucleosis	a viral disease characterized by enlarged lymph nodes, an increase in number of mononuclear blood cells (monocytes and lymphocytes), sore throat, and fatigue

Necrosis	the death of one or more cells or a portion of a tissue or organ
Nosocomial infection	a disorder, usually bacterial infections, contracted during a hospital stay, often due to antibiotic-resistant strains of *Staphylococcus*
Pernicious anemia	an anemia caused by an inadequate supply of folic acid (vitamin B_{12}), resulting in red blood cells that are large, varied in shape, and reduced in number
Plague	any infectious disease of wide prevalence or excessive mortality; it also refers specifically to an acute infectious disease caused by the bacterium *Yersinia pestis* and characterized by high fever, skin eruptions, internal hemorrhage, and pneumonia; also called *bubonic plague*
Rabies	a bacterial infection spread from the mouth of an infected animal, usually by way of a bite; this bacterium produces a neurotoxin that acts on the central nervous system and is highly fatal
Septicemia	a systemic disease caused by the presence of bacteria and their toxins in the circulating blood: a person suffering from this is referred to as "septic"
Sickle cell anemia	an inherited, chronic anemia that is characterized by defective hemoglobin that causes red blood cells to become misshapen (sickle-shaped), resulting in drowsiness, leg ulcerations, fever, joint and abdominal pain, and thrombosis

Smallpox	viral disease caused by the *variola* virus
Staphylococcemia	the presence of *Staphylococci* bacteria in the blood, which is the literal meaning of the term; commonly called a staph infection, it is a frequent complication to normal healing and also the most common cause of food poisoning, skin inflammation, osteomyelitis, and nosocomial infections
Streptococcemia	presence of the bacterium *Streptococcus* in the blood
Tetanus	a disease caused by a powerful neurotoxin released by the common bacterium *Clostridium tetani;* the toxin acts on the central nervous system to cause convulsions and paralysis
Thymoma	a tumor originating in the thymus gland

Treatments, Procedures, and Devices

Prefix	Definition	Combining Form	Definition	Suffix	Definition
anti-	*against or opposite of*	aden/o	*gland*	-crit	*to separate*
pro-	*before*	aut/o	*self*	-ectomy	*surgical excision of removal*
		bi/o	*life*	-ic	*pertaining to*

globin/o	*protein*	-logous	*pertaining to study*
hem/o, hemat/o	*blood*	-logy	*study of*
		-lysis	*loosen or*
hom/o	*same*	-stasis	*standing still*
immun/o	*exempt or immunity*	-therapy	*treatment*
lymph/o	*clear water or fluid*	-tic	*pertaining to*
thromb/o	*clot*	-phylaxis	*protection*

Medical Term	**Definition**
Antibiotic therapy	a therapeutic treatment in which a substance with known toxicity to bacteria, which can be obtained from mold (fungus) or from other bacteria, is administered; it is effective only against bacteria, many types of which are capable of developing resistance, especially when antibiotics are not administered properly
Anticoagulant	a chemical agent that inhibits the clotting process
Antiretroviral therapy	the application of drugs to battle against a class of therapy viruses that tend to mutate quickly, known as *retroviruses,* of which HIV is a

member; also known as *combination therapy,* the drugs form a

cocktail that includes nucleoside analog reverse transcriptase

inhibitors and protease inhibitors, all of which block HIV replication

by a variety of means

Attenuation   a process in which pathogens are rendered less virulent, prior to their

incorporation into a vaccine preparation

Autologous transfusion   transfusion of blood donated by a patient for personal use; this is a

common procedure before a surgery to avoid potential

incompatibility or contamination

Blood chemistry   a test or series of tests on plasma to measure the levels of particular

components (glucose, albumin, cholesterol, etc.)

Blood culture   a test to determine infection in the blood by placing a blood sample

on a nutritive media in an effort to grow populations of bacteria for

analysis

Blood transfusion   the introduction of blood, blood products, or blood substitute into a

patient's circulation to restore blood volume to normal levels; the

two main types of blood transfusions are *autologous transfusion* and

homologous transfusion

Coagulation time   a timed blood test to determine the time required for a blood clot to

form; one type of this test, called *prothrombin time (PT),* measures

the time required for prothrombin, a precursor protein, to form

thrombin and is often used to monitor anticlotting therapy; another type of test is *partial thromboplastin time (PTT),* which is used to evaluate clotting ability

Complete blood count — a common laboratory blood test that provides diagnostic information about a patient's general health; it includes several more specific tests, including *hematocrit, hemoglobin, red blood count,* and *white blood count*

Differential count — microscope count of the number of each type of white blood cell, using a stained blood smear

Hematocrit — a test that measures the percentage of red blood cells in a volume of blood, it is obtained from centrifuging a sample of blood to separate blood cells

Hematology — the general field of medicine focusing on blood-related disease

Hemoglobin — a test that measures the level of hemoglobin in red blood cells

Hemostasis — stoppage of bleeding

Homologous transfusion — transfusion of blood that is voluntarily donated by another person; it requires blood type matching known as cross-matching to prevent incompatibility

Immunization — a procedure that provides immunity against a particular antigen

Immunology — the study concerned with immunity and allergy

Immunotherapy — used in the treatment of infectious disease, it is the use of agents *(serum, gamma globulin, treated antibodies, etc.)* to activate or strengthen the immune response

Lymphadenectomy	excision of a lymph node
Platelet count	calculation of the number of platelets in the blood
Prophylaxis	any treatment that tends to prevent the onset of an infection or other type of disease
Red blood count	measures the number of red blood cells per cubic centimeter
Splenectomy	excision of the spleen
Thrombolysis	the process of dissolving a blood clot
Thymectomy	excision of the thymus gland
Vaccination	the inoculation of a culture that has reduced virulence as a means of providing a cure or a prophylaxis
Vaccine	any preparation used to activate an immune response

Abbreviations

Abbreviation	**Meaning**
AIDS	acquired immune deficiency syndrome
CBC	complete blood count
HCT, Hct	hematocrit
HGB, Hgb	hemoglobin

HIV	human immunodeficiency virus
PLT	platelet count
PT	prothrombin time
PTT	partial thromboplastin time
RBC	red blood cell or red blood count
WBC	white blood cell or white blood count

Factoid: The two most common blood types in the U.S. population are O+ (37.4%) and A+ (35.7%).

Factoid: Cytotoxic T cells help rid the body of cells that have been infected by viruses, as well as cells that have been transformed by cancer. They are also responsible for the rejection of tissue and organ grafts.

Factoid: The thymus lies in the thorax cavity directly behind the sternum. It plays an important role in the body's immune system. In times of great stress or grief, humans worldwide have automatically clutched their upper chests or "thumped" on their sternum or thymus. For centuries, people of all faiths have been known to pray while "thumping" on their chests. There is a theory that suggests that the natural tendency of humans is to "thymus thump," or stimulate the thymus when in distress, to evoke the natural immune response of the body.

CHAPTER SEVEN

Blood and the Lymphatic System

Worksheet 1

Phonetic Spelling Challenge: Spell the medical term correctly in the space provided.

1. vih RALL oh jee   _____

2. IM yoo NALL oh jee   _____

3. HEE mah TALL oh jist   _____

4. loo koh PEE nee ah   _____

5. EYE a troh JEN ik dis EEZ   _____

6. sep tih SEE mee ah   _____

7. splee NEK toh mee   _____

8. hee MAT oh krit   _____

9. HIGH droh FOH bee ah   _____

10. in FEK shun   _____

Spelling Challenge: These terms are spelled incorrectly. Spell each term correctly in the space provided.

1. Splean   _____

2. Anemmia   _____

3. Botellism   _____

4. Hemorhagic   _____

5. Hemoglobeinopathy   _____

6. Allergey   _____

7. Fungemea   _____

8. Noseocomial   _____

9. Tettanous   _____

10. Anterretroviral   _____

11. Lymfadeenectomy   _____

12. Hemgloobin   _____

13. Thimoma   _____

14. Streptosoccemia   _____

15. Molaria   _____

Abbreviation Matchup: Select and match the correct abbreviation to the definition.

_____ **1.** hematocrit

_____ **2.** white blood cell

_____ **3.** partial thromboplastin time

_____ **4.** complete blood count

_____ **5.** human immunodeficiency virus

_____ **6.** prothrombin time

_____ **7.** hemoglobin

_____ **8.** red blood cell

_____ **9.** platelet count

   a. CBC

   b. HGB, Hgb

   c. HCT, Hct

   d. RBC

   e. WBC

   f. PTT

   g. PT

   h. PLT

   i. HIV

True/False: Mark each statement as true (T) or false (F).

_____ **1.** Lymph transports substances throughout the body, but this fluid is found only within lymphatic vessels.

_____ **2.** Lymph is formed from blood during capillary exchange and rejoins the bloodstream later.

_____ **3.** Both blood and lymph carry white blood cells.

_____ **4.** An abnormally reduced number of white blood cells in a sample of blood is a symptom of disease called hemorrhage.

_____ **5.** The presence of abnormally large red blood cells in a sample of blood is a sign of disease and is called leukopenia.

_____ **6.** Abnormal enlargement of the spleen is a symptom of injury or infection.

_____ **7.** A mass of blood outside of the blood vessels and confined within an organ or space within the body, usually in a clotted form, is called hematoma.

_____ **8.** A disease caused by a parasitic protozoan that infects red blood cells and the liver during different parts of its life cycle is called malaria.

_____ **9.** Inflammation of the lymph nodes is a condition called lymphadenitis.

_____ **10.** A form of cancer of lymphatic tissue that is characterized by the progressive enlargement of lymph nodes, fatigue, and deficiency of the immune response is called Hodgkin's disease.

Fill in the Blank: Fill in the blank with the correct medical term from this chapter.

11. An infectious disease that causes internal bleeding (hemorrhage) and high fevers is generally known as _____.

12. A fungal infection that spreads throughout the body by way of the bloodstream is called _____.

13. _____ means "bad temperament."

14. A(n) _____ is the body's immune response to allergens, which are antigens that produce a hypersensitivity reaction, including immediate inflammation.

15. The presence of toxins in the bloodstream is a symptom known as

_____.

16. The presence of bacteria in a sample of blood is a sign of an infection and is called

_____.

17. The presence of large, irregularly shaped red blood cells in a sample of blood is

called _____.

18. An abnormal increase in the number of red blood cells in the blood is called

_____.

19. Abnormal enlargement of the spleen is a symptom of injury or infection and is called

_____.

20. The rupture of the red blood cell membrane is called

_____.

Short Answer

Write the definitions for each of the following terms.

21. Anaphylaxis _____

22. Diphtheria _____

23. Gas gangrene _____

24. Hemophilia _____

25. Iatrogenic disease _____

CHAPTER EIGHT

THE CARDIOVASCULAR SYSTEM

LEARNING OBJECTIVES

After completing this chapter, students will be able to:

- Define and spell the word parts used to create terms for the cardiovascular system.

- Break down and define common medical terms used for symptoms, diseases, disorders, procedures, treatments, and devices associated with the cardiovascular system.

- Build medical terms from the word parts associated with the cardiovascular system.

- Pronounce and spell common medical terms associated with the cardiovascular system.

INSTRUCTIONAL GOAL: DEFINE AND SPELL THE WORD PARTS USED TO CREATE MEDICAL TERMS FOR THE CARDIOVASCULAR SYSTEM.

Content Abstract

The two major parts of the cardiovascular system

1. The heart—hollow muscular pump located in the thoracic cavity

2. Blood vessels—transport blood throughout the body

Other parts of the cardiovascular system

1. Pericardial sac, or parietal pericardium—the thick, parchmentlike membrane that

covers the heart

2. Pericardial cavity—the potential space enclosed by the pericardial sac that is nearly filled by the presence of the heart

3. Epicardium, or visceral pericardium—a thin membrane that forms the outer surface of the heart

4. Myocardium—cardiac muscle

5. Endocardium—the thin inner lining of the heart

6. Atria—two thin-walled upper chambers that receive incoming blood to the heart

7. Ventricles—two thick-walled lower chambers that push blood out of the heart

8. Vena cavae—two large veins that deposit blood into the right atrium

9. Coronary sinus—drains the heart into the right atrium

10. Valves:

- Atrioventricular valves

  - Mitral (bicuspid) valve—located on the left side of the heart, between the atrium and ventricle

  - Tricuspid valve—located on the right side of the heart, between the atrium and ventricle

- Semilunar valves

  - Pulmonary semilunar valve—located between the right ventricle and the pulmonary artery, which carries blood to the lungs

- Aortic semilunar valve—located between the left ventricle and the aorta, the largest vessel in the body; carries blood to every part of the body with the exception of the lungs

Three circulations:

1. Pulmonary circulation—blood associated with the lungs

   - Pulmonary trunk—located on the right side of the heart and carries blood from the right ventricle to the lungs. These arteries then branch extensively to form a vast capillary network. Here the blood gases are exchanged between the capillaries and air sacs of the lungs.

2. Systemic circulation—the distribution of the aortic branches and the return of blood to the heart by way of veins

   - Aorta—the largest vessel in the body carries oxygenated blood from the left ventricle to every part of the body except the lungs.

   - Cardiac output—the volume of blood that enters the systemic circulation with each heartbeat

3. Coronary circulation—consists of vessels that arise from a small branch at the base of the aorta to supply the heart wall.

A collection of specialized cells that generate or carry impulses that stimulate the heart to contract include the:

1. Sinoatrial (SA) node

   - A cluster of cells in the right atrium that serve as the pacemaker of the heart

2. Atrioventricular (AV) node

- A second cluster of cells that relay the signal to the ventricles after a brief delay

3. Atrioventricular (AV) bundle, or bundle of His

- Cells that form a conduction pathway through the walls of the ventricles

Pronunciation for medical terms in this chapter can be found:

- In the text in parentheses following the term

- In the audio glossary on the DVD-ROM

- In the audio glossary at the Companion Website

Combining Form	Definition
angi/o	*blood vessel*
aort/o	*aorta*
arter/o, arteri/o	*artery*
atri/o	*atrium*
cardi/o	*heart*
coron/o	*crown or circle, heart*
my/o, myos/o	*muscle*
pector/o	*chest*
valvul/o	*little valve*
vas/o	*blood vessel*

vascul/o	*little blood vessel*
ven/o	*vein*
ventricul/o	*little belly, ventricle*

Teaching Strategies

- Say each new term in class, and have the students repeat.

- Stress the importance of using instructional aids to practice pronunciation.

- Encourage students to learn more about cardiovascular professionals by visiting the following Websites: Alliance of Cardiovascular Professionals at www.acp-online.org; American Society of Echocardiography at www.asecho.org.

Practical Activity

The heart makes two distinct sounds, referred to as "lubb-dupp." These sounds are produced by the forceful snapping shut of the heart valves. "Lubb" is the closing of the atrioventricular valves. "Dupp" is the closing of the semilunar valves. Listening to these sounds with a stethoscope is part of determining if the valves are functioning properly.

Factoid

By the end of a long life, a person's heart can have beaten (expanded and contracted) more than 3.5 billion times. In fact, each day, the average heart beats 100,000 times, pumping about 2,000 gallons (7,571 liters) of blood.

INSTRUCTIONAL GOAL: BREAK DOWN AND DEFINE COMMON MEDICAL TERMS USED FOR SYMPTOMS, DISEASES, DISORDERS, PROCEDURES, TREATMENTS, AND DEVICES FOR THE CARDIOVASCULAR SYSTEM.

Content Abstract

Signs and Symptoms

Prefix	Definition	Combining Form	Definition	Suffix	Definition
a-	*without or absence*	angi/o	*blood vessel*	-a	*singular*
dys-	*abnormal, bad, painful, or difficult*	cardi/o	*heart*	-algia, -dynia	*condition of pain*
brady-	*slow*	cyan/o	*blue*	-genic	*pertaining to formation*
tachy-	*rapid*	pect/o, pector/o	*chest*	-ia	*condition of*
		rhythm/o, rrhythm/o	*rhythm*	-osis	*condition of*
		sten/o	*narrow*	-plegia	*paralysis*
				-sis	*state of*
				-spasm	*sudden involuntary*

Medical Term	Definition
Angina pectoris	chest pain usually caused by an insufficient supply of blood to the heart
Angiospasm	abnormal spasms of a blood vessel wall
Angiostenosis	narrowing of a blood vessel
Arrhythmia	any loss of rhythm in the heartbeat
Bradycardia	an abnormally slow heart rate, usually under 50 beats per minute
Cardiodynia	a sensation of pain in the heart
Cardiogenic	a condition originating in the heart
Cyanosis	a symptom in which a blue tinge is seen in the skin and mucous membranes, caused by oxygen deficiency
Dysrhythmia	a disturbance or abnormality of the heart's normal rhythmic cycle
Palpitation	a pounding, racing, or skipping of the heartbeat
Perfusion deficit	a reduction of blood flow through a vessel, which can be caused by an occlusion or narrowing
Tachycardia	a fast heartbeat

Diseases and Disorders

Prefix	Definition	Combining Forms	Definition	Suffix	Definition
endo-	*within*	angi/o	*blood vessel*	-ac	*pertaining to*
epi-	*upon, above, on top, over*	aort/o	*aorta*	-ade	*process*
hyper-	*excessive, abnormally high, above*	arter/o, arteri/o	*artery*	-al	*pertaining to*
hypo-	*deficient, abnormally low, below*	ather/o	*fatty*	-ary	*pertaining to*
para-	*beside, departure from normal*	atri/o	*atrium*	-emia	*condition of blood*
peri-	*around*	cardi/o	*heart*	-ic	*pertaining to*
poly-	*many*	coron/o	*crown or circle, heart*	-ion	*process of*
		isch/o	*hold back*	-itis	*inflammation*
		my/o	*muscle*	-megaly	*large*

127

scler/o	*thick or hard; sclera*	-oma	*tumor*
sept/o	*wall or partition; putrefying*	-osis	*condition of*
sten/o	*narrow*	-pathy	*disease*
tampon/o	*plug*		
tens/o	*pressure*		
valvul/o	*little valve*		
varic/o	*dilated vein*		
ventricul/o	*little belly, ventricle*		

Medical Term	Definition
Aneurysm	the bulging of an arterial wall caused by a congenital defect or an acquired weakness of the arterial wall produced as blood is pushed against it
Angiocarditis	inflammation of the heart and blood vessels
Angioma	a tumor arising from a blood vessel
Aortic insufficiency	when the aortic valve fails to close completely during diastole, blood may return to the left ventricle, causing the left ventricle to work harder

Aortic stenosis	a narrowing of the aorta that reduces the flow of blood
Aoritis	inflammation of the aorta
Arteriopathy	a general term for a disease of an artery
Arteriosclerosis	hardening of the arteries in which the artery walls lose their elasticity and become brittle
Atherosclerosis	narrowing of an artery due to the deposition of a fatty plaque along the internal wall
Atrial septal defect	a congenital condition characterized by an opening (ASD) in the septum that separates the right and left atria, allowing blood to pass between them
Atriomegaly	the artria are enlarged or dilated, reducing their ability to push blood into the ventricles
Atrioventricular	a defect, usually congenital, that alters the structure defect of both an atrium and a ventricle
Cardiac arrest	cessation of heart activity
Cardiac tamponade	acute compression of the heart due to the accumulation of fluid within the pericardial cavity
Cardiomegaly	abnormal enlargement of the heart
Cardiomyopathy	a general disease of the heart muscle
Cardiovalvulitis	inflammation of the heart valves
Coarctation of the	a congenital disease in which the aorta is narrowed, causing reduced

aorta	systemic circulation and fluid accumulation in the lungs
Congestive heart failure (CHF)	a chronic condition characterized by the inability of the left ventricle to pump enough blood through the body to adequately supply systemic tissues; also called *left ventricular failure*
Cor pulmonale	literally "heart lung" in French, this chronic enlargement of the right ventricle results from congestion within the pulmonary circulation; also called *right ventricular failure*
Coronary artery disease (CAD)	a generalized condition of the arteries of the heart characterized by a reduction of blood flow to the heart wall, for which the most common cause is atherosclerosis
Coronary occlusion	blockage of an artery supplying the heart, often due to atherosclerosis
Embolism	a blood clot or foreign particle that moves through the circulation. It can produce severe circulatory restriction when it becomes lodged in an artery (pl = emboli).
Endocarditis	inflammation of the endocardium; when it is caused by a bacterial infection, it is known as *bacterial endocarditis*
Fibrillation	uncoordinated, rapid contractions of the ventricles or atria, resulting in circulatory collapse
Heart block	an interference with the normal electrical conduction of the heart, often the result of an MI affecting SA or AV nodes

Heart murmur	an abnormal, soft blowing or rasping sound heard through auscultation of the heart
Hemorrhoid	a varicose vein of the anal region that produces symptoms of local pain and itching
Hypertension	persistently high blood pressure, which includes essential hypertension, where the condition is not traceable to a single cause, and secondary hypertension, where the high blood pressure is caused by the effects of another disease, such as atherosclerosis
Hypotension	a chronic condition of low blood pressure
Ischemia	an abnormal low flow of blood to tissue
Myocardial infarction (MI)	the death of a portion of the myocardium, usually caused by an occluded vessel interrupting blood flow
Myocarditis	inflammation of the myocardium, or muscle layer of the heart wall
Patent ductus arteriosus	a congenital condition characterized by an opening between the pulmonary artery and the aorta, allowing blood to pass across; in this condition, the connecting channel that is a normal part of fetal circulation before birth fails to close
Pericarditis	inflammation of the pericardium, usually affecting both layers
Phlebitis	inflammation of a vein
Polyarteritis	inflammation of an artery at numerous sites

Septicemia   a bacterial infection of the bloodstream

Tetralogy of Fallot   four congenital defects associated with the heart—pulmonary stenosis, ventricular septal defect, incorrect position of the aorta, and right ventricular hypertrophy—combined; as a result, the pulmonary circulation is bypassed

Thrombosis   the presence of stationary blood clots within blood vessels

Varicosis   a condition of an abnormally dilated vein

Ventricular septal defect   an opening in the septum that separates the right ventricle from the left ventricle, allowing blood to pass between them

Treatments, Procedures, and Devices

Prefix	Definition	Combining Form	Definition	Suffix	Definition
endo-	*within*	angi/o	*blood vessel*	-ac	*pertaining to*
		aort/o	*aorta*	-ary	*pertaining to*
		arter/o, arteri/o	*artery*	-ectomy	*surgical excision or removal*
		cardi/o	*heart*	-gram	*a record or image*

coron/o	*crown or circle, heart*	-graphy	*measurement, recording process*
ech/o	*sound*	-lytic	*to loosen or dissolve*
electr/o	*electricity*	-meter	*measure*
embol/o	*plug*	-metry	*process of measuring*
man/o	*gas*	-plasty	*surgical repair*
phleb/o	*vein*	-scopy	*process of viewing*
pulmon/o	*lung*	-stomy	*surgical opening*
son/o	*sound*	-tomy	*incision or to cut*
sphygm/o	*heartbeat*		
thromb/o	*clot*		
valvul/o	*little valve*		

Medical Term	Definition
Angiogram	a recording obtained from an angiography procedure, it is an X-ray of a blood vessels after injection of a contrast medium
Angioplasty	general surgical repair of a blood vessel; it includes procedures to reopen blocked vessels
Angioscopy	use of a flexible fiber-optic instrument, or endoscope, to observe a diseased blood vessel to assess the lesion and decide on a mode of treatment; the procedure also includes use of a camera, video recorder, and monitor
Angiostomy	the surgical procedure that involves the creation of an opening into a blood vessel
Angiotomy	the surgical incision into a blood vessel
Aortography	a procedure that obtains an X-ray image, MRI, or CAT scan image of the aorta.
Aortogram	a recording of an X-ray of the aorta
Arteriogram	a recording of an X-ray of a particular artery
Arteriotomy	an incision into an artery
Arterioplasty	procedure to repair an injured artery
Arteriorrhaphy	suturing an opening in an artery
Auscultation	physical exam consisting of listening to internal sounds using a stethoscope; sounds that suggest abnormalities are often caused by dysrhythmias

Cardiac catheterization	insertion of a narrow, flexible tube, or catheter, through a coronary blood vessel to withdraw blood samples, measure pressure, and inject contrast medium for imaging purposes
Cardiac pacemaker	a battery-powered device implanted under the skin and wired to the SA node; it produces timed electric pulses that replace the pacemaking function of the SA node
Cardiopulmonary resuscitation	an emergency response procedure (CPR) that includes artificial ventilation and external heart massage
Coronary artery bypass graft	surgical procedure (CABG) in which a blood vessel is removed from another part of the body and inserted in the coronary circulation to bypass blood flow around an occluded artery
Coronary stent	a plastic scaffold that is used to anchor a surgical implantation; in this case, it is implanted in a coronary artery to prevent closure of the artery after angioplasty or atherectomy
Defibrillation	electrical charge to the heart in an effort to defibrillate, or to stop fibrillation of, the heart, delivered by paddles onto the skin of the chest, or to the heart muscle directly, if the chest has been opened
Doppler sonography	an ultrasound procedure that checks blood flow in an effort to determine the causes of a localized reduction in blood flow
Echocardiography	an ultrasound procedure where sound waves are directed through the heart to evaluate heart anomalies—the recorded data are called an echocardiogram; if

performed during exercise to identify heart conditions, it is called a stress echocardiogram, or *stress ECHO*

Electrocardiography	procedure (ECG or EKG) in which the electrical events associated with the beating of the heart are evaluated, represented by deflections of a pen or graph known as an *electrocardiogram*: when it is measured during physical activity using a treadmill or ergonometer, it is called a stress *electrocardiogram* and is useful in detecting heart conditions
Embolectomy	surgical removal of a floating blood clot, or embolus
Endarterectomy	incision into an artery, usually to remove fatty plaque or a blood clot
Holter amblutory monitor	a portable electrocardiograph worn by the patient, which monitors electrical activity of the heart over 24-hour periods, proving useful in detecting periodic or transient abnormalities
Nitroglycerin	a drug that is commonly used as an emergency vasodilator
Phlebectomy	excision of a vein
Phlebotomy	incision into a vein, usually to remove blood for sampling or to donate blood; a technician who performs this procedure is called a *phlebotomist*
Positron emission tomography	scan or procedure (PET) that provides blood flow images using techniques with radioactive isotope labeling
Sphygmomanometry	procedure that measures arterial blood pressure using a device called a sphygmomanometer
Thrombolytic	treatments dissolving blood clots using drugs such as streptokinase or tissue

therapy plasminogen activator; this treatment is often applied within six hours of an MI

Valvuloplasty surgical repair of a heart valve; if repair is not possible, a valve replacement might be required using an artificial valve or a porcine valve

Factoids

- Sudden cardiac death from coronary heart disease occurs over 680 times per day in the United States.

- Cellular phones can interfere with pacemaker functioning. Digital phones are more likely to cause problems than analog phones. Keeping the cellular phone at least six inches away from the pacemaker could decrease the chances of problems. Issues regarding cellular phones should be discussed with the patient's heart doctor.

- In cities where defibrillation is provided within 5 to 7 minutes, the survival rate from cardiac arrest is as high as 49%.

Factoid

African Americans of both sexes and Caucasian males have higher rates of significant hypertension. Although essential hypertension has no correctable cause, some genetic factors have been identified.

Abbreviations

Abbreviation	Meaning
ASHD	atherosclerotic heart disease
ASD	atrial septal defect
AV	atrioventricular
CABG	coronary artery bypass graft
CAD	coronary artery disease
CHF	congestive heart failure
CPR	cardiopulmonary resuscitation
ECG, EKG	electrocardiogram
MI	myocardial infarction
PET	position emission tomography

Factoid

Porcine valves—artificial valves made from the aortic valves of pigs—are used in valve replacement surgery. An advantage to porcine valves is that patients only have to take anticoagulant medication for 6 to 8 weeks after the surgery. The disadvantage is that porcine valves do wear out and usually must be replaced about 10 years after the replacement surgery.

CHAPTER EIGHT

The Cardiovascular System

Worksheet 1

Phonetic Spelling Challenge

Spell the medical term correctly in the space provided.

1. ah RITH mee ah   _____

2. ahr TEE ree oh skleh ROH siss   _____

3. kor pull moh NAY lee   _____

4. HIGH poh TEN shun   _____

5. throm BOH siss   _____

6. NIGH troh GLIH ser ihn   _____

7. dee fib rih LAY shun   _____

8. AN jee oh plass tee   _____

9. fleh BOT oh mee   _____

10. ahr tee ree OG rah fee   _____

Spelling Challenge: These terms are spelled incorrectly. Spell each term correctly in the space provided.

1. Dopler sonographey   _____

2. Anjiotomy   _____

3. Halter monitor   _____

4. Phebectomey   _____

5. Cardioemeggaly   _____

6. Arteriopathey   _____

7. Myoncarditis   _____

8. Valveoloplasty   _____

9. Phebottomy   _____

10. Spygmomanometry   _____

Abbreviation Matchup: Select and match the correct abbreviation to the definition.

_____   **1.** positron emission tomography   **a.**   AV

_____   **2.** congestive heart failure   **b.**   ASHD

_____   **3.** myocardial infarction   **c.**   CABG

_____   **4.** atrioventricular   **d.**   CHF

_____   **5.** coronary artery disease   **e.**   CPR

_____   **6.** atherosclerotic heart disease   **f.**   MI

_____   **7.** atrial septal defect   **g.**   ECG/EKG

_____   **8.** coronary artery bypass graft   **h.**   PET

_____   **9.** electrocardiogram   **i.**   ASD

_____   **10.** cardiopulmonary resuscitation   **j.**   CAD

True/False: Mark each statement as true (T) or false (F).

_____ **1.** A heart attack is an MI.

_____ **2.** Inflammation of the membrane surrounding the heart is called phlebitis.

_____ **3.** An abnormally low flow of blood to tissues is the condition known as hypotension.

_____ **4.** An inflammation the valves of the heart is called cardiovalvulitis.

_____ **5.** An injury to the atrioventricular node (AV node), which normally receives impulses from the sinoatrial node (SA node) and transmits them to the ventricles to stimulate ventricular systole, is called an AV block.

Fill in the Blank: Fill in the blank with the correct medical term from this chapter.

11. The presence of stationary blood clots within one or more blood vessels is called _____.

12. Death of a portion of the myocardium is called _____.

13. An abnormally dilated vein is called _____.

14. A congenital disease characterized by aortic stenosis that is present at birth is known as _____.

15. A(n) _____ is a general term that means "blockage."

16. Persistently high blood pressure is an abnormal condition called _____.

17. The use of a flexible fiberoptic instrument, or endoscope, to observe a diseased blood vessel in order to assess the lesion is a procedure called

_____.

18. A procedure that obtains an X-ray image of an artery is known as

_____.

19. An incision into an artery is called _____.

20. Insertion of a narrow, flexible tube, called a catheter, through a coronary vessel into the heart is called _____.

Short Answer: Write the definition for each of the following terms.

21. Cardiac pacemaker   _____

22. CABG   _____

23. Echocardiography   _____

24. Embolectomy   _____

25. Endarterectomy   _____

CHAPTER NINE

THE RESPIRATORY SYSTEM

LEARNING OBJECTIVES

After completing this chapter, students will be able to:

- Define and spell the word parts used to create terms for the respiratory system.

- Break down and define common medical terms used for symptoms, diseases, disorders, procedures, treatments, and devices associated with the respiratory system.

- Build medical terms from the word parts associated with the respiratory system.

- Pronounce and spell common medical terms associated with the respiratory system.

INSTRUCTIONAL GOAL: DEFINE AND SPELL THE WORD PARTS USED TO CREATE MEDICAL TERMS FOR THE RESPIRATORY SYSTEM.

Content Abstract

The *respiratory system* brings oxygen into the bloodstream, which can then transport it to all body cells. The process is known as *respiration*. Besides bringing oxygen to the bloodstream, the respiratory system also removes the waste product carbon dioxide from the blood and channels it outside the body. There are four steps to respiration:

- Inhalation, or inspiration—the movement of air from the outside environment to the tiny air sacs, alveoli, within the lungs

- External respiration—occurs when fresh air has filled the lungs and the air molecules diffuse between the alveoli and the capillaries, and carbon dioxide moves in the opposite direction.

- Internal respiration—occurs when oxygen carried in the bloodstream diffuses into surrounding body cells, and carbon dioxide moves from the cells into the bloodstream.

- Exhalation, or expiration—pushing of the used air and carbon dioxide out of the body. Together with inhalation, this is known as ventilation.

The organs of the respiratory system include:

1. The conducting portion—the chambers and tubes that conduct air that extends from the nose to the lungs. They are all lined with mucous membranes and cilia, which:

   - Serve to warm and humidify the air on the way to the lungs.

   - Trap foreign particles.

   - Form a conveyer belt of motion that transports the foreign particles to the mouth or nose for elimination when you cough or sneeze. The expelled mucus is known as sputum.

The organs referred to as the upper respiratory tract are as follows:

- Nose—within the nose is the nasal cavity. Other portions include:

   - Nasal septum—the central partition, which divides the nasal cavity.

   - Paranasal sinuses—spaces within the bones of the face and skull, which are connected to the nasal cavity

- Pharynx, or throat—surrounded by muscles, and serves as a common chamber for swallowing food and breathing air. Inhaled air enters the larynx, while swallowed air enters the esophagus.

- Larynx, or voice box—the structure that produces sound when exhaled air is squeezed between folds of membrane that partially block the airway. Important parts include:

  - Glottis—the opening to the larynx

  - Epiglottis—a flap of cartilage that prevents food from entering the larynx

- Trachea or windpipe—a foot-long tube that carries air between the larynx and the bronchi. It is prevented from collapsing by the presence of stiff cartilage rings, which strengthen its walls to form a rigid tube.

The organs referred to as the lower respiratory tract are as follows:

- Bronchi—begin as two branches from the distal end of the trachea, forming the right and left primary bronchi. The walls are kept rigid by the presence of cartilage rings.

- Bronchial tree—the subdivisions of the bronchi as they branch within the lung.

- Bronchioles—thin-walled branches of the bronchial tree. Because their walls are not supported by cartilage, they can collapse due to respiratory disorders. The bronchiole leads into cul-de-sacs, each of which opens into a cluster of microscopic, saclike alveoli.

2. The respiratory portion—the alveoli form the substance of the lung. They are one

cell thick and lie adjacent to capillary walls. The barrier between them is very thin and is known as the respiratory membrane. It is here that gas diffusion between the lungs and the blood takes place. Alveoli contain a specialized type of white blood cell called alveolar macrophage, which removes inhaled foreign particles like dust, pollen, and bacteria.

- Lungs—spongy, soft organs that fill half of the thoracic cavity. Each lung is divided into lobes and further divided into smaller compartments called segments. Other portions consists of:

  - Visceral pleura—a thin, almost transparent layer of serous membrane located on the outer surface of the lung

  - Parietal pleura—serous membrane attached to the inside wall of the thorax

  - Pleural cavity—the space between the visceral and parietal pleura

Combining Forms	Definition
alveol/o	*air sac, alveolus*
bronch/o	*airway, bronchus*
hem/o, hemat/o	*blood*
laryng/o	*voice box, larynx*
lob/o	*a rounded part, lobe*
muc/o	*mucus*
nas/o	*nose*

ox/o	*oxygen*
pharyng/o	*throat, pharynx*
phragm/o, phragmat/o	*partition*
pleur/o	*pleura, rib*
pneum/o, pneumon/o, pneumat/o	*air, lung*
pulmon/o	*lung*
rhin/o	*nose*
sept/o	*wall, partition; putrefying*
sinus/o	*cavity*
thorac/o	*chest, thorax*
trache/o	*windpipe, trachea*

Practical Activity

CPR training is a wonderful addition to any allied health program. It is not reserved for caregivers, but appropriate for everyone of nearly every age. Consider taking a course in CPR for the bystander and renewing the certification at least every two years.

Factoid: The central nervous system's respiratory center is located in the lateral medulla oblongata of the brain stem.

INSTRUCTIONAL GOAL: BREAK DOWN AND DEFINE COMMON MEDICAL TERMS USED FOR SYMPTOMS, DISEASES, DISORDERS, PROCEDURES, TREATMENTS, AND DEVICES FOR THE RESPIRATORY SYSTEM.

Content Abstract

Signs and Symptoms

Prefix	Definition	Combining Forms	Definition	Suffix	Definition
a-, an-	*without or absence of*	bronch/o	*airway*	-algia	*condition of pain*
brady-	*slow*	hem/o	*blood*	-capnia	*condition of carbon dioxide*
dys-	*bad, abnormal, painful, or difficult*	laryng/o	*larynx, voice box*	-dynia	*pain*
epi-	*upon, over, above, oron top*	orth/o	*straight*	-emia	*condition of blood*

eu-	*normal, good, well*	rhin/o	*nose*	-oxia	*condition of oxygen*
hyper-	*excessive, above normal, or above*	thorac/o	*chest, thorax*	-phonia	*condition of sound or voice*
hypo-	*deficient, below normal, or below*			-pnea	*breath*
tachy-	*rapid or fast*			-ptysis	*to cough up*
				-rrhagia	*condition of bleeding, hemorrhage*
				-spasm	*sudden involuntary muscle contraction*
				-staxis	*dripping*

Medical Term	Definition
Acapnia	absence of carbon dioxide
Anoxia	absence of oxygen
Aphonia	absence of voice
Apnea	inability to breathe
Bradypnea	slow breathing
Bronchospasm	narrowing of the airway caused by contraction of smooth muscles in the wall of the bronchioles
Cheyne-Stokes respiration	a pattern of breathing marked by a gradual increase of deep breathing, followed by shallow breathing that leads to apnea
Dysphonia	hoarseness of the voice
Dyspnea	difficulty breathing
Epistaxis	a nosebleed
Eupnea	normal breathing
Hemoptysis	coughing up and spitting out blood originating from the lungs
Hemothorax	blood in the pleural cavity
Hypercapnia	excessive carbon dioxide in the blood

Hyperpnea	deep breathing
Hyperventilation	excessive movement of the air in and out of the lungs
Hypocapnia	deficient levels of carbon dioxide in the blood
Hypopnea	shallow breathing
Hypoventilation	a breathing rhythm that fails to meet the body's gas exchange demands
Hypoxemia	deficient levels of oxygen in the blood
Hypoxia	deficient levels of oxygen in tissues throughout the body
Laryngospasm	spasmodic closure of the glottis
Orthopnea	the ability to breathe is limited to when in an upright position
Paroxysm	a sudden sharp pain or convulsion
Sputum	expectorated matter, usually containing mucus and sometimes pus
Tachypnea	rapid breathing
Thoracalgia	pain in the chest region

Diseases and Disorders

Prefix	Definition	Combining Forms	Definition	Suffix	Definition
a-	*without or absence of*	atel/o	*incomplete*	-al, -ic	*pertaining to*
epi-	*upon, over, above, on top*	bronch/o, bronchi/o	*airway*	-ectasis	*dilation or expansion*
		carcin/o	*cancer*	-genic	*pertaining to producing, formation, or causing*
		coni/I	*dust*	-ia, -ism, -osis	*condition of*
		cyst/o	*bladder*	-itis	*inflammation*
		embol/o	*throwing in*	-oma	*tumor*
		fibr/o	*fiber*		
		glott/o	*opening into the windpipe*		
		laryng/o	*larynx, voice box*		

myc/o	*fungus*
nas/o	*nose*
pharyng/o	*throat, pharynx*
pleur/o	*pleura, rib*
pneum/o,	*lung or air*
pneumon/o	
pulmon/o	*lung*
py/o	*pus*
rhin/o	*nose*
sinus/o	*cavity*
sphyx/o	*pulse*
sten/o	*narrowing*
thorac/o	*chest, thorax*
tonsill/o	*almond*
trache/o	*trachea*
tubercul/o	*small swelling*

Medical Term	Definition
Asphyxia	the absence of respiratory ventilation or suffocation
Asthma	a condition of the lungs characterized by widespread narrowing of the bronchioles and formation of mucus plugs, producing fits of wheezing, shortness of breath, and coughing; it is caused by the local release of factors during an allergic response
Atelectasis	the absence of gas in the lungs due to a failure of alveolar expansion; also called *collapsed lung*
Bronchiectasis	dilation of the bronchi
Bronchitis	inflammation of the bronchi
Bronchogenic carcinoma	cancer originating in the bronchi
Bronchopneumonia	acute inflammation of the smaller bronchial tubes, bronchioles, and alveoli
Chronic obstructive pumonary disease	a group of disorders associated with the obstruction of bronchial airflow, usually as a result of inhaling tobacco products. The disorders include *emphysema, chronic bronchitis,* and *bronchospasm.*
Coccidioidomycosis	a fungal infection of the upper respiratory tract and lungs that often spreads to other organs; also known as *valley fever,* it is caused by inhaling dust containing

spores of *Coccidioides immitis*

Coryza	a common viral head cold
Croup	a disease of infants and young children, caused by acute obstruction of the larynx and characterized by a hoarse cough
Cystic fibrosis	a hereditary disease characterized by excess mucous production in the respiratory tract and elsewhere
Emphysema	a chronic lung disease characterized by enlarged alveoli and a damaged respiratory membrane; its symptoms include apnea, a barrel chest due to labored breathing, and gradual deterioration due to chronic hypoxemia
Epiglottitis	inflammation of the epiglottis
Laryngitis	inflammation of the larynx
Legionellosis	a form of pneumonia caused by the bacterium *Legionella pneumophelia,* also called *Legionnaires' disease*
Nasopharyngitis	inflammation of the nose and pharynx
Pertussis	an acute infectious disease characterized by inflammation of the larynx, trachea, and bronchi, producing spasmodic coughing; also called *whooping*

cough because of the noise produced during coughing when the larynx spasms

Pharyngitis	inflammation of the pharynx
Pleural effusion	escape of fluid into the pleural cavity
Pleuritis	inflammation of the pleurae; also called *pleurisy*
Pneumoconiosis	inflammation of the lungs caused by the chronic inhalation of fine particles, which leads to the formation of a fibrotic tissue around the alveoli that reduces their ability to stretch; it includes *asbestos inhalation* and silicosis caused by fine silicone dust inhalation
Pneumonia	inflammation of soft lung tissue, excluding the bronchi, caused by bacteria, viral, or fungal infection, in which the alveoli become filled with fluid
Pneumonitis	inflammation of the lungs independent of a particular cause
Pneumothorax	presence of air or gas in the pleural cavity
Pulmonary edema	accumulation of fluid in the alveoli and bronchioles
Pulmonary embolism	blockage in the pulmonary circulation caused by a moving blood clot
Pyothorax	a condition of pus in the pleural cavity; also called

empyema

Respiratory distress syndrome	respiratory failure characterized by atelectasis, also called *hyaline membrane disease;* this condition occurs in two forms. Neonatal respiratory distress syndrome appears in infants and is caused by insufficient surfactant cells; adult respiratory distress syndrome affects adults and is caused by severe lung infection or injury.
Rhinitis	inflammation of the nasal mucous membrane
Sinusitis	inflammation of the sinus mucous membranes
Tonsillitis	inflammation of a tonsil, usually a palatine tonsil; an inflamed pharyngeal tonsil is called an adenoid
Tracheitis	inflammation of the trachea
Tracheostenosis	narrowing of the trachea
Tuberculosis	infection of the lungs by the bacterium *Mycobacterium tuberculosis*
Upper respiration infection	infection of the upper respiratory tract; usually the result of a virus

Treatments, Procedures, and Devices

Prefix	Definition	Combining Forms	Definition	Suffix	Definition
anti-	*against or opposite of*	aden/o	*gland*	-al	*pertaining to*
endo-	*within*	angi/o	*blood vessel*	-ary	*pertaining to*
		bronch/o	*airway*	-centesis	*surgical puncture*
		dilat/o	*to widen*	-ectomy	*surgical removal or excision*
		laryng/o	*voice box, larynx*	-gram	*a record or image*
		lob/o	*round part, lobe*	-graphy	*measurement, recording process*
		ox/I	*oxygen*	-ion	*process*
		pleur/o	*pleura, rib*	-meter	*measuring device*
		pneum/o,	*Lung, airway*pneum	-metry	*measurement*

158

on/o			
pulmon/o	*lung*	-oid	*resembling*
rhin/o	*nose*	-plasty	*surgical repair*
spir/o	*breathe*	-scopy	*process of viewing*
thorac/o	*chest, thorax*	-stomy	*surgical creation of an opening*
trache/o	*windpipe, trachea*	-tomy	*incision or to cut*

Medical Term	Definition
Acid-fast bacilli smear	a clinical test performed on sputum to identify the presence of bacteria that reacts to acid, including *Mycobacterium tuberculosis*
Adenoidectomy	excision of a swollen pharyngeal tonsil, known as an adenoid
Antihistamine	a therapeutic drug that inhibits the effects of histamines, which are compounds released by cells that cause bronchial constriction and blood vessel dilation

Arterial blood gases	a clinical test on arterial blood to identify the levels of oxygen and carbon dioxide
Aspiration	the removal of fluid with suction
Auscultation	a physical examination that listens to sound within the body, often with the aid of a stethoscope
Bronchodilation	use of a bronchodilating agent in an inhaler to reduce bronchial constriction and thereby improve breathing
Bronchogram	X-ray image of the bronchi
Bronchography	the process of obtaining an X-ray of the bronchi
Bronchoscopy	bronchi are examined with a *bronchoscope,* a modified type of endoscope
Chest CT scan	diagnostic imaging of the chest by a CT scanning instrument; used to diagnose respiratory tumors, scan pleural effusion, pleurisy, and other diseases by providing three-dimensional imaging
Chest X-ray	an X-ray photograph of the thoracic cavity used to diagnose tuberculosis, tumors, and other lung conditions; also called a *chest radiograph*
Ear, nose, and throat specialist	a physician specializing in the treatment of upper respiratory diseases
Endoscopy	visual examination of a body space with the use of an

instrument with a flexible tube that contains mirrors or a camera, called an *endoscope;* it is a noninvasive technique for diagnostic and treatment purposes

Endotracheal intubation   insertion of a tube into the trachea via the nose or mouth to open the airway

Expectorant   a drug that breaks up mucus and promotes coughing to remove the mucus

Incentive spirometry   a postoperative breathing therapy in which a portable spirometer is used by a patient to encourage lung exercise; it reduces pulmonary complications

Laryngectomy   surgical removal or excision of the larynx

Laryngoscopy   procedure that examines the larynx with a *laryngoscope*

Laryngostomy   surgical creation of an opening into the larynx

Laryngotracheotomy   incision into the larynx and trachea

Lobectomy   excision of a section or lobe of a lung

Mechanical ventilation   a technique used by a respiratory therapist or EMT to provide assisted breathing using a *ventilator,* which pushes air into the patient's airway

Nebulizer   a device used to convert a liquid medication to a mist and deliver it to the lungs with the aid of deep inhalation

Oximetry   measurement of oxygen levels in the blood using an instrument called an *oximeter;* a *pulse oximeter* is a noninvasive procedure using an oximeter that is pressed against the fingertip

Pleurocentesis	surgical puncture and aspiration of fluid from the pleural cavity
Pneumonectomy	excision of a lung
Pulmonary angiography	X-ray of the blood vessels of the lungs following injection of a contrast medium
Pulmonary function test	diagnostic test performed to determine the cause of lung disease by evaluating lung capacity through the use of spirometry; types include *tidal volume,* which is the amount of air expired after a normal expiration, and *vital capacity,* which is the amount of air exhaled after a maximal exposure
Pulmonologist	a physician specializing in the treatment of lung disease
Resuscitation	artificial respiration used to restore breathing; the most common form is cardiopulmonary resuscitation
Rhinoplasty	surgical repair of the nose
TB skin test	a test to determine the presence of a TB infection, in which a purified protein derivative sample of the TB bacillus is injected intradermally; also called *PPD skin*

	test or *Mantoux* skin test
Thoracocentesis	surgical puncture into the chest cavity to aspirate; also called *thoracentesis*
Thoracoscopy	examination of the thoracic cavity using a *thoracoscope*
Thoracostomy	surgical puncture into the chest cavity, usually for the insertion of a tube; the procedure is often termed "placing a chest tube"
Thoracotomy	incision into the chest
Tonsillectomy	excision of one or more tonsils, usually palatine
Tracheoplasty	surgical repair of the trachea
Tracheostomy	surgical creation of an opening into the trachea, usually for the insertion of a tube
Tracheotomy	incision into the trachea
Ventilation-perfusion scanning	a diagnostic tool of nuclear medicine that is used to evaluate pulmonary function, it can identify pulmonary embolism and pulmonary edema; it is also called *lung scan* and *V/P scan*

Factoids

- Non-small-cell lung cancer accounts for about 80% of all lung cancer cases.

- Cor pulmonale causes about 25% of all types of heart failure.

- An estimated 8,000 to 18,000 people get Legionnaires' disease in the United States each year. Some people can be infected with the *Legionella* bacterium and have mild symptoms or no illness at all.

Factoids

- A pneumothorax can occur in three forms: spontaneous pneumothorax, trauma pneumothorax, and tension pneumothorax.

- Chest X-ray is the most commonly performed diagnostic X-ray examination.

- Approximately half of all X-rays obtained in medical institutions are chest X-rays.

Abbreviations.

Abbreviation	Meaning
ABGs	arterial blood gases
ARDS	adult (acute) respiratory distress syndrome
AFB	acid-fast bacilli
CF	cystic fibrosis
COPD	chronic obstructive pulmonary disease
CPR	cardiopulmonary resuscitation
CXR	chest X-ray
HMD	hyaline membrane disease
LTB	laryngotracheobronchitis

NRDS   neonatal respiratory distress syndrome

PE   pulmonary embolism

PPD   purified protein derivative

RDS   respiratory distress syndrome

SARS   severe acute respiratory syndrome

TB   tuberculosis

URI   upper respiratory infection

VPS or V/Q scan ventilation-perfusion scanning

Factoids

- The primary instrument used in pulmonary function testing is the spirometer. It is designed to measure changes in volume and can only measure lung volume compartments that exchange gas with the atmosphere.

- Both lungs have about 750 million alveoli.

- If all of the alveoli were laid out flat and stitched together like a patchwork quilt, it would be the size of a tennis court.

CHAPTER NINE

The Respiratory System

Worksheet 1

Phonetic Spelling Challenge: Spell the medical term correctly in the space provided.

1. at eh LEK tah siss   _____

2. kroop   _____

3. NOO moh THOH raks   _____

4. too BER kyoo LOH siss   _____

5. loh BEK toh mee   _____

6. rye NYE tiss   _____

7. PLOO ral eh FYOO zhun   _____

8. ep ih glah TYE tiss   _____

9. SISS tik fye BROH siss   _____

10. brong KYE tiss   _____

Spelling Challenge: These terms are spelled incorrectly. Spell each term correctly in the space provided.

1. Respirration   _____

2. Dysfonia   _____

3. Kine-Strokes respiration   _____

4. Hyperpnia _____

5. Laringospasm _____

6. Asma _____

7. Emfisema _____

8. Bronkitis _____

9. Appnea _____

10. Bronchiactasis _____

Abbreviation Matchup: Select and match the correct abbreviation to the definition.

_____ **1.** pulmonary embolism

_____ **2.** cystic fibrosis

_____ **3.** chest X-ray

_____ **4.** cardiopulmonary

       resuscitation

_____ **5.** tuberculosis

a. TB

b. AFB

c. NRDS

d. PE

e. CPR

_____ **6.** acid-fast bacilli

_____ **7.** neonatal respiratory distress

_____ **8.** upper respiratory infection

_____ **9.** arterial blood gases

_____ **10.** chronic   obstructive

       pulmonary disease

f. ABGs

g. CXR

h. COPD

i. URI

j. CF

True/False: Mark each statement as true (T) or false (F).

_____ **1.** A reduced breathing rhythm that fails to meet the body's gas exchange demands is called hyperventilation.

_____ **2.** Abnormally shallow breathing is called hypopnea.

_____ **3.** A nosebleed is clinically called hemoptysis.

_____ **4.** A narrowing of the airway by the contraction of smooth muscles in the walls of the tiny tubes within the lungs known as bronchioles is called bradypnea.

_____ **5.** The common term for atelectasis is collapsed lung.

Fill in the Blank: Fill in the blank with the correct medical term from this chapter.

11. Bronchogenic carcinoma is commonly referred to as

_____.

12. This hereditary lung disease strikes 1 in roughly 2,500 children and is commonly fatal due to lung destruction before the age of 30 years.

_____

13. The symptoms of this lung disease arise when the alveoli lose their elasticity, causing them to burst, which reduces the efficiency of gas exchange.

_____

14. Inflammation of the nose and pharynx is called _____.

15. The most common cause of adenocarcinoma of the lung is

_____.

Short Answer: Write the definition for each of the following terms.

21. Pneumonia   _____

22. Pyothorax   _____

23. Tracheitis   _____

24. Adenoidectomy   _____

25. Aspiration   _____

CHAPTER TEN

THE DIGESTIVE SYSTEM

LEARNING OBJECTIVES

After completing this chapter, students will be able to:

- Define and spell the word parts used to create terms for the digestive system.

- Break down and define common medical terms used for symptoms, diseases, disorders, procedures, treatments, and devices associated with the digestive system.

- Build medical terms from the word parts associated with the digestive system.

- Pronounce and spell common medical terms associated with the digestive system.

INSTRUCTIONAL GOAL: DEFINE AND SPELL THE WORD PARTS USED TO CREATE MEDICAL TERMS FOR THE URINARY SYSTEM.

Content Abstract

The *digestive system* converts food into a form that the body can use for energy, growth, and repair. It derives its name from its primary function, *digestion.* Accompanying the function of digestion are secondary functions, which include:

- Ingestion—the introduction of food and drink

- Mastication—the process of chewing food in the mouth

- Swallowing—a muscular process that moves food through the throat

- Absorption—the transport of nutrients and water molecules from the digestive tract to the bloodstream

- Defecation—the release of solid waste, or *feces*

The digestive system is divided into two main parts:

1. The gastrointestinal (GI) tract or alimentary canal:

   - Mouth or oral cavity—where the processing of food begins. It consists of:

   - Palate—the roof of the oral cavity, which includes:

     - Hard palate—the anterior, bony portion

     - Soft palate—the posterior, boneless portion, which ends as a projection that hangs downward like a grape, known as the *uvula*

   - The floor of the oral cavity consists of:

     - Tongue—assists in swallowing and in speech

     - Teeth—border the front and sides of the oral cavity, providing hard surfaces for *mastication*

   - Pharynx—the tube from the internal nares to the esophagus (and larynx). Movement of food is achieved by the swallowing reflex, which pushes food into the esophagus.

   - Esophagus—a muscle-walled tube that carries food from the pharynx to the stomach. It is lined with a protective *mucosa.* The food is moved to the stomach by muscular contractions known as *peristalsis,* which occur through the GI tract.

The distal end enters the abdominal cavity at an opening through the *diaphragm,* which is lined with a membrane called the *peritoneum.*

- Stomach—where *chemical* digestion of proteins takes place. The stomach is lined with a mucosa, which protects the stomach lining from the digestive enzymes. It also contains many one-celled glands, which secrete a mixture known as *gastric juice,* made up of *hydrochloric acid (HCl);* a protein called pepsinogen (converts to *pepsin,* a protein-cleaving enzyme, in the environment created by HCl); and mucus. The stomach is divided into four regions:

  - Cardia—located near the esophagus

  - Fundus—the upper domed part

  - Body—the central stomach

  - Pylorus—the lower area that joins the small intestine; the ring of muscle that borders the stomach and small intestine is the *pyloric valve*

- Small intestine—a coiled tube, about 20 feet in length, that extends between the stomach and the large intestine. This is the site for chemical digestion completion. Each section is lined with a mucosa that includes tiny fingerlike projections, known as villi. The villi improve the small intestine's ability to absorb nutrients, which is its primary function. Peristaltic contractions of the small intestine move food slowly to maximize absorption, eventually pushing the watery, indigestible waste into the large intestine. The small intestine includes:

  - Duodenum—first portion

  - Jejunum—second portion

- Ileum—final portion

- Large intestine—forms the waste into a solid material, known as *feces,* by absorbing water as it moves the material in slow, peristaltic contractions. The three sections are:

  - Cecum—this short, pouchlike section receives material from the small intestine. It includes a 2- to 5-inch-long dead-ended appendage known as the *appendix.*

  - Colon—extends from the cecum for about 8 feet and includes:

    - Ascending segment

    - Transverse segment

    - Descending segment

    - Sigmoid segment

  - Rectum—the terminal end of the large intestine, which opens to the body's exterior by way of the anus

2. The accessory organs—the supportive organs that lie outside the GI tract. Each organ manufactures material that benefits the process of digestion. They consist of:

- Salivary glands—begin the process of chemical digestion in the mouth by secreting *saliva.* The saliva enters the oral cavity through small ducts and includes an enzyme called *amylase* that breaks down *polysaccharides* to begin the process of chemical digestion. There are three pairs:

  - Parotid glands—the largest pair, is located in the cheeks

- Submandibular glands—located behind the chin

- Sublingual glands—located below the tongue

- Liver—the largest visceral organ of the body, has cells, known as *hepatic cells,* that perform many functions, including:

  - Bile production—Bile is a greenish fluid delivered to the small intestine by way of *biliary ducts,* where it aids in the digestion of fat. The main duct is the *common bile duct.* It gets its color from pigments resulting from the recycling process of RBCs.

  - The interconversion of nutrients

  - The recycling of red blood cell components

  - The removal of toxins from the blood

- Gallbladder—a small sac located in a small depression on the posterior side of the liver. It receives bile from the liver, which it stores until a meal is ingested.

- Pancreas—an oblong organ located behind the stomach. Most of its cells produce digestive enzymes, which are delivered to the small intestine by way of ducts. The enzymes, collectively called *pancreatic juice,* enable the small intestine to complete the process of chemical digestion. Other cells of the pancreas secrete the hormones *insulin* and *glucagons,* which regulate sugar levels in the blood.

Combining Form	Definition
abdomin/o	*abdomen, abdominal cavity*
an/o	*anus*
append/o, appendic/o	*appendix*
bil/I	*bile*
cec/o	*blind intestine, cecum*
chol/e	*bile, gall*
choledoch/o	*common bile duct*
col/o, colon/o	*colon*
cyst/o	*bladder*
dent/o	*teeth*
duoden/o	*twelve, duodenum*
enter/o	*small intestine*
esophag/o	*gullet, esophagus*
gastr/o	*stomach*
gingiv/o	*gums*
gloss/o	*tongue*
hepat/o	*liver*
ile/o	*to roll, ileum*

jejun/o	*empty, jejunum*
lingu/o	*tongue*
or/o	*mouth*
pancreat/o	*sweetbread, pancreas*
peps/o, pept/o	*digestion*
periton/o	*stretch over, peritoneum*
pylor/o	*pylorus*
rect/o	*rectum*
sial/o	*saliva*
sigm/o	*the letter S, sigmoid*
stomat/o	*mouth*

Factoid

A major cause of peptic ulcer, although far less common than *H. pylori* or NSAIDs, is Zollinger-Ellison syndrome. A large amount of excess acid is produced in response to the overproduction of the hormone gastrin, which in turn is caused by tumors on the pancreas or duodenum. These tumors are usually malignant and must be removed and acid production suppressed to relieve the recurrence of the ulcers.

INSTRUCTIONAL GOAL: BREAK DOWN AND DEFINE COMMON MEDICAL TERMS USED FOR SYMPTOMS, DISEASES, DISORDERS, PROCEDURES, TREATMENTS, AND DEVICES FOR THE DIGESTIVE SYSTEM.

Content Abstract

Signs and Symptoms

Prefix	Definition	Combining Forms	Definition	Suffix	Definition
a-	*without or absence of*	bil/I	*bile*	-algia	*condition of pain*
dia-	*through*	flux/o	*flow*	-dynia	*pain*
dys-	*bad, painful, difficult, or abnormal*	gastr/o	*stomach*	-emesis	*vomiting*
				-emia	*condition of blood*
re-	*back*	halit/o	*breath*	-ia	*condition of*
		hemat/o	*blood*	-megaly	*abnormally large*
		hepat/o	*liver*	-osis	*condition of*

peps/o, pept/o	*digestion*	-rrhea	*excessive discharge*
phag/o	*eat, sw allow*		
steat/o	*fat*		

Medical Term	**Definition**
Aphagia	inability to swallow
Ascites	an accumulation of fluid within the peritoneal cavity; a symptom of liver dysfunction
Constipation	reduced peristalsis in the large intestine, resulting in infrequent or incomplete defecation
Diarrhea	frequent discharge of watery fecal material, which can be caused by an improper diet, but more commonly is caused by infection of virus, bacteria, or protozoa; it can lead to severe dehydration
Dyspepsia	indigestion
Dysphagia	difficulty in swallowing
Flatus	a condition of gas trapped in the GI tract or released through the anus
Gastrodynia	pain in the stomach
Halitosis	bad breath
Hematemesis	vomiting blood

Hepatomegaly   enlargement of the liver

Jaundice   a yellowish staining of the skin, sclera of the eyes, and deeper tissues caused

                 by the accumulation of bile pigment in the bloodstream that normally is

                 removed by the liver, and thus a symptom of liver dysfunction; it can also be a

                 symptom of red blood cell destruction

Nausea   from the Latin and Greek word for "seasickness," it is a symptomatic urge to

                 vomit; when accompanied by vomiting, it may be abbreviated *N & V.*

Reflux   a backward flow of material in the GI tract, or regurgitation

Steatorrhea   abnormal fat levels in the feces

Diseases and Disorders

Prefix	Definition	Combining Forms	Definition	Suffix	Definition
an	*not*	aden/o	*gland*	-al	*pertaining to*
dys	*bad, painful, abnormal, or difficult*	appendic/o	*appendix*	-ectasis	*dilation or expansion of*
mal	*bad*	cheil/o	*lip*	-ia	*condition*
		chol/e	*bile, gall*	-ic	*pertaining to*
		choledoch/o	*common bile*	-it is	*inflammation*

179

	duct		
cirrh/o	*orange*	-malacia	*softening*
col/o	*colon*	-megaly	*abnormally large*
cyst/o	*bladder, sac*	-oid	*resembling*
diverticul/o	*diverticulum*	-oma	*tumor*
duoden/o	*twelve, duodenum*	-osis	*condition of*
enter/o	*small intestine*	-pathy	*disease*
esophag/o	*esophagus*	-penia	*abnormal reduction in number or deficiency*
gastr/o	*stomach*	-ptosis	*drooping*
gingiv/o	*gums*	-sis	*state of*
gloss/o	*tongue*		
hem/o	*blood*		
hepat/o	*liver*		
lip/o	*fat*		
lith/o	*stone*		

orex/o	*appetite*
pancreat/o	*pancreas*
parot/o	*parotid gland*
pept/o	*digestion*
periton/o	*peritoneum*
polyp/o	*small growth*
proct/o	*anus*
rect/o	*rectum*
volv/o	*to roll*

Medical Term	**Definition**
Anorexia nervosa	a personality disorder characterized by an extreme aversion to food that results in weight loss and malnourishment
Appendicitis	inflammation of the appendix
Bulimia	an eating disorder involving repeated gorging with food that is followed by induced vomiting or laxative abuse; commonly known as "binging and purging"
Cheilitis	inflammation of the lip
Choledocholithiasis	presence of mineralized masses, called gallstones, or stones, in the

common bile duct where they block bile flow

Cholelithiasis	generalized condition of gallstones
Cirrhosis	a chronic, progressive liver disease resulting from hepatic cell failure, which can be caused by chronic alcoholism or viral infection
Colitis	inflammation of the colon; when the condition is chronic and results in the formation of colonic ulcers, it is called *ulcerative colitis,* the main symptom of which is severe bloody diarrhea
Colorectal cancer	cancer of the colon and rectum, which often originates as a polyp and becomes an aggressive, metastatic tumor
Crohn's disease	chronic inflammation of any part of the GI tract, most commonly the ileum, that involves small ulcerations of the intestinal wall, resulting in scar tissue formation and intestinal obstruction; also called *regional ileitis* or *regional enteritis,* it is usually an inherited condition
Diverticulitis	inflammation of abnormal small pouches in the wall of the colon
Diverticulosis	condition of diverticula in the colon
Duodenal ulcer	an ulcer at the wall of the duodenum
Dysentery	severe inflammation of the intestine marked by frequent diarrhea, abdominal pain, fever, and dehydration; it is usually caused by infection by bacteria or protozoa

Enteritis	inflammation of the small intestine
Esophagitis	inflammation of the esophagus
Gastrectasis	abnormal stretching or dilation of the stomach
Gastric cancer	cancer of the stomach, also called *stomach carcinoma*
Gastric ulcer	an ulcer of the wall of the stomach
Gastritis	inflammation of the stomach
Gastroenteritis	inflammation of the stomach and small intestine
Gastroenterocolitis	inflammation of the stomach, small intestine, and colon
Gastroesophageal	recurring backflow of stomach contents into the esophagus as a result of a weakened lower esophageal sphincter, producing burning pain
Gastromalacia	softening of the stomach wall
Giardiasis	infection of the intestinal tract by the protozoa *Giardia intestinalis* or *G. lambia,* producing symptoms of diarrhea, cramps, nausea, and vomiting
Gingivitis	inflammation of the gums
Glossitis	inflammation of the tongue
Glossopathy	generalized disease of the tongue
Hemorrhoids	a varicose condition of veins in the anus that results in sometimes painful swellings

Hepatitis	inflammation of the liver
Hepatoma	a tumor of the liver
Hiatal hernia	protrusion of part of the stomach upward through an opening in the diaphragm normally penetrated by the esophagus, known as the *esophageal hiatus*
Inflammatory bowel disease	a syndrome affecting the intestines and characterized by a wide range of symptoms and conditions, ranging from periodic diarrhea and flatus to ulcerative colitis and Crohn's disease
Inguinal hernia	protrusion of a loop of the small intestine through the abdominal wall; when occurring among males, it can be a protrusion into the scrotal cavity
Intussusception	an infolding of a segment of the intestine within another segment
Irritable bowel syndrome	a chronic disease characterized by periodic disturbances of large intestinal function, such as diarrhea and constipation, without clear physical damage; attacks are characterized by abdominal pain caused by accumulation of gas and abdominal muscle spasm
Lactose intolerance	a lack of the enzyme lactase in the small intestine, producing symptoms of gas production and diarrhea when dairy foods are consumed
Pancreatitis	inflammation of the pancreas
Parotitis	inflammation of the parotid gland; also called *mumps*

Peptic ulcer	an erosion in the wall of the stomach, duodenum, or any other part of the GI tract that might be exposed to gastric juice that usually is due to a reduction of the protective mucous layer; about 80% of peptic ulcers are correlated to an infection from *Helicobacter pylori*
Peritonitis	inflammation of the peritoneum
Polyp	any abnormal mass of tissue that projects outward from a wall; they are usually benign growths that can occur in the nose, throat, and intestines
Polyposis	presence of many polyps, usually in the colon or rectum, that poses a great risk for malignancy
Proctitis	inflammation of the rectum and anus
Proctoptosis	prolapse of the rectum
Strangulated hernia	a hernia that is constricted, which reduces blood flow to the organ; if early intervention does not occur, the organ can develop gangrene
Umbilical hernia	a protrusion of a loop of the intestine through the abdominal wall in the umbilical region
Volvulus	a twisting of the intestine that leads to obstruction

Treatments, Procedures, and Devices

Prefix	Definition	Combining Form	Definition	Suffix	Definition
an	*without or absence of*	abdomin/o	*abdomen, abdominal cavity*	-al, ic	*pertaining to*
anti	*against*	acid/o	*a solution or substance with a pH less than 7*	-centesis	*surgical puncture*
endo	*within*	append/o	*appendix*	-ectomy	*surgical removal*
		cheil/o	*lip*	-emetic	*pertaining to vomiting*
		cholecyst/o	*gallbladder*	-gram	*a record or image*
		choledoch/o	*common bile duct*	-graphy	*measurement or recording process*
		col/o	*colon*	-plasty	*surgical repair*

fec/o	*feces*	-rrhaphy	*suturing*
gastr/o	*stomach*	-scopy	*to view*
gingiv/o	*gums*	-spasmodic	*pertaining to a sudden, involuntary contraction*
gloss/o	*tongue*	-stomy	*surgical creation of an opening*
ile/o	*to roll, ileum*	-tomy	*incision or to cut*
lapar/o	*abdomen*		
lith/o	*stone*		
polyp/o	*small growth*		
pylor/o	*pylorus*		
vag/o	*vagus nerve*		

Medical Term	Definition
Abdominocentesis	a surgical puncture through the abdominal wall to remove fluid; also called *paracentesis*
Antacid	a drug that neutralizes stomach acid
Antiemetic	a drug that prevents or stops vomiting
Antispasmodic	a drug that decreases peristalsis in the GI tract to arrest spasm or diarrhea
Appendectomy	surgical removal, or excision, of the appendix
Barium enema	an enema containing barium administered for a *lower GI series* diagnostic test
Cathartic	a drug that stimulates peristalsis of the colon; also called a *laxative*
Cheilorrhaphy	suture of the lip
Cholecystectomy	excision of the gallbladder
Cholecystogram	X-ray image of the gallbladder, which is used to confirm diagnosis of cholelithiasis
Choledocholithotomy	incision into the common bile duct, which is performed to remove one or more obstructive stones
Colectomy	excision of the colon
Colostomy	surgical creation of an opening into the colon by way of the abdominal wall, which establishes an artificial anus and can be

temporary or permanent as a treatment for cancer, obstructions, or ulcerative colitis

Fecal occult blood test	a lab test performed to detect blood in the feces
Gastrectomy	surgical removal of any part of the stomach or, in extreme cases, the entire organ
Gavage	the process of feeding a patient through a tube inserted into the nose that drops into the stomach, called a *nasogastric tube*
GI endoscopy	visual examination of the GI tract using an endoscope, which includes a camera, fiber-optics, and long flexible tube; endoscopic procedures utilized in GI tract diagnostics include the *colonoscopy, esophagogastroduodenoscopy, esophagoscopy, gastroscopy, laparoscopy, proctoscopy,* and *sigmoidoscopy*
Gingivectomy	surgical removal of diseased tissue in the gums
Glossorrhaphy	suture of the tongue
Hemorrhoidectomy	excision of hemorrhoids
Herniorrhaphy	surgical repair of a hernia
Ileostomy	surgical opening through the abdominal wall and into the ileum to establish a secondary anus for the passage of feces
Laparotomy	incision into the abdomen
Polypectomy	excision of a polyp

Pyloroplasty	surgical repair of the pylorus region of the stomach or the pyloric valve
Stool culture and sensitivity	collection of a fecal sample and of the growth of microorganisms from it in a culture to identify a pathogenic cause of disease
Upper GI series	diagnostic X-ray images of the stomach and duodenum following the administration of barium as a radiopaque contrast medium
Vagotomy	surgical dissection of branches of the vagus nerve, which innervates much of the GI tract; it is performed to reduce gastric juice secretion to treat chronic gastric ulcer

Factoids

- Lactose intolerance is genetic and occurs most often in people of African, Asian, and Mediterranean descent.

- Individuals affected with polyposis have approximately a 10% lifetime risk of developing cancer in the duodenum, a cancer that is very rare in the general population.

Factoid

Uvulectomy is a traditional surgery performed on infants and children throughout Africa and in some Middle Eastern countries. It is performed as a traditional treatment to prevent throat infections and normally is done early in infancy or in the first or second year of life.

Abbreviations.

Abbreviation	Meaning
BE	barium enema
EGD	esophagogastroduodenoscopy
FOBT	fecal occult blood test
GERD	gastroesophageal reflux disease
GI	gastrointestinal
IBD	inflammatory bowel disease
IBS	irritable bowel syndrome
LGI	lower GI series
N&V	nausea and vomiting
SCS	stool culture and sensitivity
UGI	upper GI series

Factoid

Many people who think they have a throat problem or chronic postnasal drip are surprised to discover they have GERD. An examination of the throat by a specialist is important to make the correct diagnosis.

CHAPTER 10

Digestive System

Worksheet 1

Phonetic Spelling Challenge

Spell the medical term correctly in the space provided.

1. kon stih PAY shun

   _____

2. koh LYE tiss

   _____

3. EHN ter EYE tiss

   _____

4. PROK top TOH siss

   _____

5. an tye ee MEH tik

   _____

6. GAS troh EN ter oh koh LYE tiss

   _____

7. pahr oh TYE tiss

   _____

8. gas TREK tah siss

   _____

9. dye ah REE ah

   _____

Spelling Challenge

These terms are spelled incorrectly. Spell each term correctly in the space provided.

1. Heppatoma

   _____

2. Proctopatosis

   _____

3. Vulvulus

   _____

4. Gastroesophogeal

_____

5. Hiatalle hernia

_____

6. Polyposiss

_____

7. Intususception

_____

8. Parrotitis

_____

9. Bullimia

_____

10. Cholaystitis

_____

Abbreviation Matchup

Select and match the correct abbreviation to the definition.

_____ **1.** gastrointestinal   **a.** EGD

_____ **2.** inflammatory bowel disease   **b.** SCS

_____ **3.** barium enema   **c.** UGI

_____ **4.** upper GI series   **d.** FOBT

_____ **5.** nausea and vomiting   **e.** LGI

_____ **6.** esophagogastroduodenoscopy   **f.** IBD

_____ **7.** lower GI series   **g.** BE

_____ **8.** stool culture and sensitivity   **h.** GI

_____ **9.** fecal occult blood test   **i.** N&V

_____**10.** irritable bowel syndrome   **j.** IBS

True/False

Mark each statement as true (T) or false (F).

_____ **1.** Colonoscopy views the sigmoid colon.

_____ **2.** Proctoscopy views the colon.

_____ **3.** A part of the stomach may be removed to treat peptic ulcers.

_____ **4.** IBD is associated with ulcerative colitis and Crohn's disease.

_____ **5.** Dysphagia often accompanies a sore throat, although its chronic form can be a sign of oral or pharyngeal cancer.

Fill in the Blank

Fill in the blank with the correct medical term from this chapter.

11. A disorder that is characterized by difficulty absorbing one or more nutrients is called _____.

12. The procedure of suturing a lip is called _____.

13. The surgical procedure that involves an incision through the abdominal wall, often from the base of the sternum to the pubic bone, is called a(n) _____.

14. Surgical repair of the pylorus region of the stomach, which can include repair of the pyloric valve, is known as a(n) _____.

15. A drug that reduces the acidity of the stomach cavity is called a(n)

_____.

Short Answer

Write the definition for each of the following terms.

21. Cheilorrhaphy_____

22. Cathartic_____

23. Cholelithiasis_____

24. Gingivitis_____

25. Gastromalacia_____

CHAPTER ELEVEN

THE URINARY SYSTEM

LEARNING OBJECTIVES

After completing this chapter, students will be able to:

- Define and spell the word parts used to create terms for the urinary system.

- Break down and define common medical terms used for symptoms, diseases, disorders, procedures, treatments, and devices associated with the urinary system.

- Build medical terms from the word parts associated with the urinary system.

- Pronounce and spell common medical terms associated with the urinary system.

INSTRUCTIONAL GOAL: DEFINE AND SPELL THE WORD PARTS USED TO CREATE MEDICAL TERMS FOR THE URINARY SYSTEM.

The *urinary system* functions as the sanitary engineer of the body, maintaining the purity and health of the body's fluids by removing unwanted waste materials and recycling other materials. Its organs consist of:

1. Kidneys—The most important organs of this system, the kidneys are located against the posterior abdominal wall, one on each side of the body's mid-line. They are fist-sized and shaped like a kidney bean. They are covered with a protective layer of fat and a fibrous membrane. The kidneys provide the following functions:

   - Filtration of gallons of fluid from the bloodstream every day.

- Removal of metabolic wastes, toxins, excess ions, and water that leave the body as *urine,* in a function known as *excretion.* Urine—which is mostly water, excess electrolytes and metabolic waste material, urea, and ammonia—formation occurs in three stages within each of the nephrons:

  - Filtration—The filter is a very thin membrane between the wall of the glomerulus and the inner wall of the Bowman's capsule. As blood pressure pushes blood through the glomerulus, some of the blood's plasma is forced through tiny openings in the membrane, filling the Bowman's capsule with fluid.

  - Reabsorption—As fluid flows through the renal tubule, most of the water is reabsorbed back into the bloodstream.

  - Secretion—Excess electrolytes (salts in an ionic form) and other waste are transported into the renal tubule.

- Returns needed materials back to the blood.

- Regulates blood pressure, pH, and red blood cell production in bone marrow.

Parts of the kidneys are as follows:

- Hilum—the concave margin where the renal artery, renal veins, nerves, and *ureter* join the kidney

- Renal pelvis—a membrane-lined basin that collects urine in the center

- Renal medulla—lies external to the pelvis

- Cortex—the outermost area of the kidney

- Nephrons—located in the renal medulla and cortex, they amount to about one million in number. Each nephron has the following:

  - Bowman's capsule—a tube that consists of a hollow ball at one end

  - Renal tubule—a long twisted portion at the other end

  - Glomerulus—a tightly coiled capillary located within the Bowman's capsule. This structure, together with the Bowman's capsule, makes up the *renal corpuscle*.

2. Ureters—a pair of spaghetti-sized organs that transport urine from the kidneys to the *urinary bladder*

   - They arise from the renal pelvis and extend downward along each side of the vertebral column until they unite with the bladder.

   - Their walls consist of:

     - The inner mucous membrane—protects the ureter from potentially damaging effects of urine.

     - The outer layer of smooth muscle—provides waves of peristalsis that help propel the urine along its way to the urinary bladder.

3. Urinary bladder—a hollow, muscular organ located at the floor of the pelvic cavity that temporarily stores urine. The interior of the urinary bladder has openings for the two ureters and the single urethra. These three openings form a triangular region known as the *trigone,* which is a frequent site of urinary infection. The bladder is lined with an elastic mucous membrane, and the outer wall is composed of

involuntary muscle that contracts during urination.

4. Urethra—a muscular tube that drains urine from the floor of the urinary bladder and transports it to the exterior. The urethra is lined internally with a protective mucous membrane. The release of urine through the urethra is called micturition or voiding. The urethra consists of the:

- Internal urethral sphincter—a thickening of muscle at the junction between the bladder and the urethra

- External urethral sphincter—muscle that is under voluntary control that surrounds the urethra at the point where it extends through the floor of the pelvic cavity

- External urethral orifice or urinary meatus—the opening to the exterior or outside the body.

The urethra differs between males and females:

- Females—about 3 to 4 cm (1.5 inches) long and bound to the vaginal wall by connective tissue. The urinary meatus lies between the vagina and the clitoris.

- Males—about 20 cm (8 inches) long and extends from the urinary bladder to the end of the penis, where it opens as the urinary meatus. As it leaves the bladder, the male urethra passes through the prostate gland that surrounds it.

Combining Form	Definition
albumin/o	*protein*
blast/o	*developing cell*

glomerul/o	*little ball, glomerulus*
gluc/o, glyc/o, glycos/o	*sweet, sugar*
meat/o	*opening, passage*
nephr/o	*kidney*
pyel/o	*renal pelvis*
ren/o	*kidney*
ureter/o	*ureter*
urethr/o	*urethra*
ur/o, urin/o	*urine*

Factoid

Urethral cancer is an extremely rare form of cancer (only about 700 cases reported worldwide). Cancer develops in the urethra as the result of abnormal cell growth within the urethra.

INSTRUCTIONAL GOAL: BREAK DOWN AND DEFINE COMMON MEDICAL TERMS USED FOR SYMPTOMS, DISEASES, DISORDERS, PROCEDURES, TREATMENTS, AND DEVICES FOR THE REPRODUCTIVE SYSTEM.

Content Abstract

Signs and Symptoms

Prefix	Definition	Combining Form	Definition	Suffix	Definition
an-	*without or absence of*	albumin/o	*albumin (a protein)*	-emia	*condition of blood*
dia-	*through*	azot/o	*urea or nitrogen*	-urea	*urine*
dys-	*bad, painful, difficult or abnormal*	bacteri/o	*bacteria*	-uresis	*urination*
poly-	*excessive, over, or many*	glycos/o	*sweet or sugar*	-uria	*pertaining to urine or urination*
		hem/o, hemat/o	*blood*		
		ket/o, keton/o	*ketone*		
		noct/o	*night*		

olig/o	*few in number*
prote/o	*protein*
py/o	*pus*

Medical Term	Definition
Albuminuria	the presence of albumin, a protein normally found in blood, in the urine
Anuresis	the inability to pass urine
Azotemia	the presence of abnormally high urea and other nitrogenous compounds in the blood
Bacteriuria	the presence of bacteria in the urine
Diuresis	the condition of passing urine (usually refers to excessive urine discharge)
Dysuria	difficulty or pain in urination
Glycosuria	the presence of glucose in the urine
Hematuria	the presence of blood in the urine
Ketonuria	the presence of ketone bodies in the urine, which is a common sign of acidosis among patients suffering from diabetes mellitus
Nocturia	urination at night
Oliguria	reduced urination

Polyuria   excessive urination

Proteinuria   the presence of protein in the urine

Pyuria   the presence of pus (white blood cells) in the urine

Diseases and Disorders

Prefix	Definition	Combining Form	Definition	Suffix	Definition
an-	*without or absence of*	albumin/o	*albumin (a protein)*	-al	*pertaining to*
dia-	*through*	azot/o	*urea or nitrogen*	-cele	*hernia, swelling, or protrusion*
dys-	*bad, painful, difficult or abnormal*	bacter/o	*bacteria*	-emia	*condition of blood*
en-	*upon, on, over, or within*	blast/o	*germ or bud*	-ia	*condition of*
epi-	*upon, over, above, or on top*	cyst/o	*bladder*	-iasis	*condition of*
hypo-	*deficient, abnormally low, or below*	glomerul/o	*ball or glomerulus*	-ic	*pertaining to*
poly-	*excessive, over, or*	hemat/o	*blood*	-itis	*inflammation*

many

ket/o, keton/o	*ketone*	-pathy	*disease*
lith/o	*stone*	-ptosis	*falling downward*
megal/o	*abnormally large*	-sis	*state of*
nephr/o	*kidney*	-urea	*urine*
olig/o	*few in number*	-uria	*urine or urination*
prote/o	*protein*		
pyel/o	*renal pelvis*		
py/o	*pus*		
ren/o	*kidney*		
spadias/o	*rip or tear*		
ur/o	*urine*		
ureter/o	*ureter*		
urethr/o	*urethra*		

Medical Term	Definition
Cystitis	inflammation of the urinary bladder
Cystocele	protrusion of the urinary bladder
Cystolith	a stone in the urinary bladder
Epispadias	a congenital defect resulting in the urinary meatus's being positioned on the dorsal surface of the penis; in females, the meatus opens dorsal to the clitoris
Glomerulonephritis	inflammation of the glomeruli within the kidney
Hydronephrosis	the condition of water in a kidney, usually caused by obstruction and backup of urine leading to distention of the renal pelvis
Hypospadias	a congenital defect in which the urinary meatus opens on the underside of the penis; in the female, the opening is within the vagina
Incontinence	the involuntary discharge of urine, which also can refer to the inability to prevent the discharge of feces; *stress incontinence* is involuntary discharge of urine due to a cough, sneeze, or strained movement
Nephritis	inflammation of a kidney
Nephroblastoma	a tumor originating from a kidney that includes developing embryonic cells; also known as *Wilms' tumor*
Nephrolithiasis	the presence of one or more stones in a kidney
Nephroma	a tumor originating from a kidney

Nephromegaly	enlargement of a kidney
Nephroptosis	the condition of a drooped kidney position, which occurs when the kidney is no longer held in its proper position; also called *floating kidney*
Polycystic kidney	a kidney condition characterized by the presence of many polyps, resulting in the loss of functional tissue
Pyelitis	inflammation of the renal pelvis
Pyelonephritis	inflammation of the renal pelvis and nephrons
Stricture	abnormal narrowing, as in urethral stricture
Uremia	an excess of urea and other nitrogenous waste in the blood, caused by failure of the kidneys to remove urea during urine formation
Ureteritis	inflammation of a ureter
Ureterocele	protrusion of a ureter
Ureterolithiasis	presence of stone(s) in a ureter
Urinary retention	abnormal accumulation of urine in the urinary bladder, resulting from an inability to urinate
Urinary suppression	an acute stoppage of urine formation by the kidneys
Urinary tract infection	infection of urinary organs—usually the urethra and bladder—in which symptoms often include fever, dysuria, and lumbar or abdominal pain

Treatments, Procedures, and Devices

Prefix	Definition	Combining Form	Definition	Suffix	Definition
a-	*without or absence of*	cyst/o, vesic/o	*bladder*	-al	*pertaining to*
dia-	*through*	hemat/o, hem/o	*blood*	-ectomy	*surgical excision or removal*
		lith/o	*stone*	-gram	*a record or image*
		meat/o	*opening*	-graphy	*recording process*
		nephr/o	*kidney*	-is	*pertaining to*
		peritone/o	*peritoneum*	-lysis	*loosen or dissolve*
		pyel/o	*renal pelvis*	-meter	*measuring instrument*
		ren/o	*kidney*	-pexy	*surgical fixation*
		son/o	*sound*	-plasty	*surgical repair*
		tom/o	*to cut*	-rrhaphy	*suturing*
		ureter/o	*ureter*	-stomy	*surgical creation of an opening*
		urethr/o	*urethra*	-tomy	*incision or to cut*
		ur/o, urin/o	*urine*	-tripsy	*surgical crushing*

Medical Term	Definition
Blood urea nitrogen	a clinical lab test that measures urea concentration in a sample of blood as an indicator of kidney function; an elevated value indicates kidney disease
Creatinine	a protein that is a normal component of urine, as a result of muscle metabolism; elevated levels in a urine indicate kidney disease
Cystectomy	excision of the urinary bladder
Cystography	an X-ray technique for imaging the urinary bladder; the resulting X-ray image is called a *cystogram*
Cystolithotomy	incision into the urinary bladder to remove a stone
Cystoplasty	surgical repair of the urinary bladder
Cystopyelography	an X-ray technique that images the urinary bladder and renal pelvis; the resulting X-ray image is called a *cystopyelogram*
Cystorrhaphy	suturing of the urinary bladder wall
Cystoscopy	use of a modified endoscope, known as a *cystoscope*, to visually examine the urinary bladder
Cystostomy	surgical creation of an artificial opening into the urinary bladder to provide an alternate exit pathway for urine
Cystotomy	incision into the urinary bladder; also called *vesicotomy*

Fulguration | a surgical procedure that destroys living tissue with an electric spark, commonly used to remove tumors or polyps from the interior bladder wall

Hemodialysis | a procedure that removes nitrogenous wastes and excess ions from the blood, replacing the normal function of the kidneys as an intervention for kidney failure, using the process of *dialysis,* in which blood is pushed through a semipermeable membrane filter to separate substances based on their molecular size

Lithotripsy | a surgical technique that crushes stones

Meatoscopy | use of a modified endoscope, known as a meatoscope, to visually examine the urinary meatus

Nephrectomy | excision of a kidney

Nephrography | an X-ray technique imaging a kidney; the X-ray image is called a *nephrogram*

Nephrolysis | freeing of the kidney from inflammatory adhesions

Nephropexy | surgical fixation of an abnormally mobile kidney

Nephroscopy | use of a modified endoscope, known as a *nephroscope*, to visually examine a kidney

Nephrosonography | an ultrasound procedure in which a kidney is imaged with the use of sound waves

Nephrostomy | surgical creation of an artificial opening into the kidney, between the renal pelvis and the kidney exterior

Nephrotomography	X-ray imaging of the kidney using sectional X-ray exposures; the image is called a *nephrotomogram*
Peritoneal dialysis	a procedure in which toxic wastes are removed from the peritoneal cavity reservoir by artificial filtration as a cleansing treatment to compensate for kidney failure
Pyelithotomy	incision into the renal pelvis to remove a stone
Pyelogram	an X-ray image of the renal pelvis; in a *retrograde pyelogram,* a contrast medium is injected into the urethra using a cystoscope, and the X-ray moves in a direction opposite from the norm, in an effort to detect the presence of stones or other obstructions; in an *intravenous pyelogram,* iodine is used as the contrast medium and is injected into the bloodstream
Pyeloplasty	surgical repair of the renal pelvis, usually involving the removal of an obstruction
Renal transplant	a surgical procedure in which a donor kidney, usually obtained from a close relative, is implanted to replace a nonfunctional kidney
Renography	a nuclear medicine test using a radioactive substance to highlight internal aspects of a kidney; the recording is called a *renogram*
Specific gravity	the relative concentration of water molecules in a liquid sample; the clinical lab test that measures specific gravity in a sample of urine evaluates filtration and water reabsorption in the kidneys using a *urinometer*

Ureterectomy	excision of a ureter
Ureterostomy	surgical creation of an artificial opening through the ureter to provide an alternate exit route for urine
Ureterotomy	incision into the wall of a ureter
Urethropexy	surgical fixation of the urethra to correct stress incontinence
Urethroplasty	surgical repair of the urethra
Urethroscopy	use of a modified endoscope, known as a *urethroscope*, to visually examine the urethra
Urethrostomy	surgical creation of an artificial opening into the urethra to establish an alternate exit route for urine
Urethrotomy	incision into the urethra
Urinalysis	a clinical lab test performed on a urine specimen, often measuring specific gravity, creatinine, glucose, protein, and pH
Urinary catheterization	insertion of a catheter, a flexible tube for channeling fluids into the urinary bladder to drain urine
Urinary endoscope	use of an endoscope to view internal structures of the urinary system; includes *cystoscopy, meatoscopy, nephroscopy,* and *urethroscopy*
Urinometer	an instrument that measures the water density of urine, a value known as specific gravity
Urologist	a physician who specializes in disorders of the urinary system

Vesicourethral suspension	a surgery performed to stabilize the urinary bladder position as a treatment of stress incontinence

Factoids

- The urinary system and the male reproductive system share some organs, particularly the urethra. The term *genitourinary* (GU) is sometimes used to describe the urinary system.

- The kidney is shaped like a kidney bean. Each weighs 4 to 6 ounces and is 2 to 3 inches wide and about 1 inch thick, or about the size of a fist. Functioning kidneys are needed for life, but it is possible to live with only one functioning kidney.

Factoid

The urinary bladder contains almost no bacteria.

Abbreviations.

Abbreviation	Meaning
BUN	blood urea nitrogen
cath	catheter, catheterization
HD	hemodialysis
IVP	intravenous pyelogram
RP	retrograde pyelogram

SG	specific gravity
UA	urinalysis
UTI	urinary tract infection
VCUG	voiding cystourethrogram

Factoids

- With age, the urinary bladder shrinks and loses some of its ability to contract and relax. As a result, older people must go to the bathroom more often.

- By 70 to 80 years of age, most people have 50% fewer nephron units and so are less able to concentrate urine.

CHAPTER 11

The Urinary System

Worksheet 1

Phonetic Spelling Challenge

Spell the medical term correctly in the space provided.

1. in KON tih nens _____

2. AL byoo men YOO ree ah _____

3. YOO rih nair ee KATH eh ter ih ZAY shun _____

4. YOO rih nair ee ehn DOSS koh pee _____

5. yoo REE throh pek see _____

6. neh FROSS koh pee _____

7. siss TOSS koh pee _____

8. yoo REE ter oh lith EYE ah siss _____

9. NEFF roh lith EYE ah siss _____

10. an yoo REE siss _____

Spelling Challenge: These terms are spelled incorrectly. Spell each term correctly in the space provided.

1. Urinarey _____

2. Diyalysis _____

3. Fullguration

_____

4. Cystonegraphy

_____

5. Urethroeplasty

_____

6. Vesicourethralle

_____

7. Urethroplexy

_____

8. Pecific gravity

_____

9. Renul

_____

10. Peritoneel

_____

Abbreviation Matchup: Select and match the correct abbreviation to the definition.

_____	**1.**	retrograde pyelogram	**a.** BUN
_____	**2.**	catheter, catheterization	**b.** IVP
_____	**3.**	intravenous pyelogram	**c.** cath
_____	**4.**	blood urea nitrogen	**d.** HD
_____	**5.**	hemodialysis	**e.** RP
_____	**6.**	voiding cystourethrogram	**f.** UTI
_____	**7.**	specific gravity	**g.** VCUG
_____	**8.**	urinalysis	**h.** UA
_____	**9.**	urinary tract infection	**i.** SG

True/False

Mark each statement as true (T) or false (F).

_____ **1.** Radioactive materials highlight internal details of the kidney.

_____ **2.** In polycystic kidney disease, polyps replace normal tissue.

_____ **3.** The involuntary release of urine, which usually occurs due to a lack of bladder control, is a condition known as enuresis.

_____ **4.** Excessive urination is a common sign of an endocrine disease.

_____ **5.** A nephroblastoma is a tumor occurring in the bladder but originating from muscle tissue.

_____ **6.** A cystocele is a stone, or calculus, in the urinary bladder.

_____ **7.** The presence of pus in the urine is called proteinuria.

_____ **8.** Reduced frequency of urination is known as nocturia.

_____ **9.** Difficulty or pain experienced during urination is a symptom of a urinary tract disease often caused by a bacterial infection.

_____**10.** The presence of ketone bodies in the urine is a sign of a kidney disorder and is called hematuria.

Fill in the Blank

Fill in the blank with the correct medical term from this chapter.

11. The presence of any protein in the urine is called _____.

12. If the exit of urine out of the kidneys becomes blocked, the urine will back up to cause distension of the renal pelvis. This painful condition is known as

_____.

13. An inflammatory condition of the renal pelvis that involves the nephrons is called

_____.

14. _____ is an incision through the urinary bladder wall.

15. The relative concentration of water molecules in a liquid sample is called

_____.

16. A combination of clinical lab tests that are performed on a urine specimen is called

a(n) _____.

17. A(n) _____ is a flexible tube that is inserted into an

opening of the body to transport fluids in or out.

Short Answer

Write the definition for each of the following terms.

21. Ureteritis _____

22. Cystoscopy _____

23. Urethropexy _____

24. Nephrosonography _____

25. Nephrolysis _____

CHAPTER TWELVE

REPRODUCTIVE SYSTEM AND OBSTETRICS

LEARNING OBJECTIVES

After completing this chapter, students will be able to:

- Define and spell the word parts used to create terms for the reproductive system and obstetrics.

- Break down and define common medical terms used for symptoms, diseases, disorders, procedures, treatments, and devices associated with the reproductive system and obstetrics.

- Build medical terms from the word parts associated with the reproductive system and obstetrics.

- Pronounce and spell common medical terms associated with the reproductive system and obstetrics.

INSTRUCTIONAL GOAL: DEFINE AND SPELL THE WORD PARTS USED TO CREATE MEDICAL TERMS FOR THE REPRODUCTIVE SYSTEM.

Content Abstract

The ***male reproductive system*** produces the sex cells, or gametes (spermatozoa or sperm cells), of the male. The male system also sustains and transports sperm cells. In addition, the male reproductive system secretes the hormone testosterone, which regulates sperm cell production and expression of secondary sex characteristics. When properly functioning, the male system has the capability to transmit sperm to a female during sexual intercourse, or coitus. The release of the sperm cell is called ejaculation, and it usually accompanies sexual climax or orgasm. The organs consists of:

1. Testes or male gonads—the primary organs of the male reproductive system. They produce male gametes (sperm cells). They initially develop within the pelvic cavity, then descend into the:

   - Scrotum—a skin-covered sac that encloses the testes. The scrotum provides an environment that is more favorable to sperm development because it has a cooler temperature than the 98°F of the pelvic cavity.

2. Male tubules—Transport sperm cells from the testes to the outside of the body. Clusters of cells between the seminiferous tubules, called *interstitial cells,* produce testosterone. Each testis is composed of about 900 of these small, tightly coiled tubules. They are the sites of sperm production. The process of sperm production is known as *spermatogenesis.* As the seminiferous tubules extend, they merge to form

larger tubules, which merge again to form a single tubule as it leaves the testes, called the *ductus epididymidis.* It coils to form an elongated structure attached to the posterior side of the testis called the *epididymis.* The ductus epididymidis terminates when it unites with a thick-walled tubule called the *vas deferens.* The walls of the vas deferens include smooth muscle, enabling the tube to form peristaltic waves that push the sperm along their route. A vas deferens extends from each testis upward through the scrotum, emerging from it to enter the pelvic cavity. The vas deferens is joined by arteries, veins, and nerves wrapped by a layer of connective tissue to form the *spermatic cord.* The vas deferens passes through the *seminal vesicles* to form the *ejaculatory duct* before uniting with the *urethra,* the final tube. It extends to its opening at the *urinary meatus* or *external urethral orifice.* The urethra passes through the *prostate gland* before entering the *penis.* Other glands at its base are called the *bulbourethral glands.*

3. Male glands—The male glands include the:

- Seminal vesicles—a pair of saclike structures at the base of the urinary bladder that secrete fluids into the vas deferens.

- Prostate gland—a single walnut-sized gland that surrounds the urethra near its emergence from the bladder and releases secretions directly into the urethra.

- Bulbourethral glands—small, pea-shaped glands located at the base of the penis, which also release secretions into the urethra. The male glands contribute to the formation of semen, which is released during ejaculation. Semen includes:

  - A fluid rich in *fructose,* for nourishing sperm cells

- A thick, milky, alkaline fluid released by the prostate gland

- A clear mucus released by the bulbourethral glands

4. Male external genitals—The external organs are the:

- Scrotum—a skin-covered sac that hangs below the pelvic cavity wall and contains the testes.

- Penis—an external cylindrical organ that contains the distal part of the urethra. It contains three cylindrical masses of erectile tissue that enable it to enlarge and harden during the physiological response known as *erection.* An erection enables the penis to be inserted into the female vagina during sexual intercourse. It consists of:

  - Glans Penis—the distal end of the penis, which is soft and spongy.

  - Prepuce—the fold of skin that covers most or all of the glans penis. It is also known as the *foreskin.* The removal of the prepuce is called *circumcision* and is usually performed soon after birth.

The ***female reproductive system*** produces the sex cells, or gametes, of the female. It also provides support for the developing embryo and fetus once fertilization has occurred and makes the process of internal fertilization possible. The female gametes are the *egg cells,* called *ova* or *oocytes.*

The ovaries and other organs are as follows:

1. The ovaries—the female gonads that produce the female gametes and the female sex hormones. The ovarian hormones are a class of compounds called *estrogens* and

progesterone. They are paired, almond-shaped organs located opposite one another against the walls of the pelvic cavity. Each one is covered by a layer of cells and is internally divided into a:

- Cortex—Contains numerous saclike ovarian *follicles,* which are in various stages of development. Each ovarian follicle contains a single ovum. A lifetime of ova are present at birth, in an immature state. Ova mature and are released on a monthly cycle called *ovulation.* The ovum bursts out of a mature *graafian follicle,* through the ovarian wall, and into the *peritoneal cavity.* From there, the ovum is usually drawn into a fallopian tube. The cycle begins at the onset of puberty and ends about 40 years later at *menopause.*

- Medulla.

2. The fallopian tubes—a pair of narrow tubes located along each side of the pelvic cavity wall; they are also called *uterine tubes* or *oviducts.* They extend approximately 10 cm (4 inches) between each ovary and the centrally located uterus and are internally lined with a ciliated mucous membrane. Near the ovary, the fallopian tube widens to form a funnel-shaped part that opens into the peritoneal cavity, called the infundibulum. The ovulated ovum is drawn into the tube by a current of mucus. If fertilization occurs, it usually happens within the upper one-third of a fallopian tube.

3. The uterus—a pear-shaped organ that is suspended just above the floor of the pelvic cavity by ligaments, located between the urinary bladder and the rectum. Parts and activities associated with the uterus include the:

- Rectouterine pouch—a narrow space that separates the uterus from the rectum

- Fundus—the upper, dome-shaped part of the uterus

- Body or corpus—the central portion of the uterus that receives the two fallopian tubes

  - Uterine cavity—the space within the body of the uterus

  - Cervix—the narrow, lower part of the uterus which includes:

  - Cervical canal—the space within the cervix

  - External os—the opening of the cervical canal into the vagina

- Perimetrium—the outer layer of the uterine wall

- Myometrium—the uterine wall's thick layer of muscle

- Endometrium—the uterine wall's inner layer rich in blood vessels, which provides an implantation site for the embryo. If implantation occurs, the inner lining of the endometrium will thicken to establish the nourishing *placenta,* which supports the developing child.

- Menstrual cycle—the increasing and decreasing in thickness of the endometrium, which occurs on a monthly cycle. It occurs approximately every 28 days. *Menses* or *menstruation* begins when the outer layer of the endometrium breaks away, causing bleeding from the uterus. The first menses, which occurs at the onset of puberty, is called *menarche.* When it ends about 40 years later, a woman is said to be in *menopause.*

4. The vagina—a thin-walled, tube-shaped organ about 8 to 10 cm (4 to 5 inches) long

that functions as a passage between the cervix of the uterus and the outside of the body. The vagina serves as:

- A passageway for menstrual blood shed by the uterus

- A passageway for semen to travel to the uterus

- A passageway for a baby to travel during birth (and is known as the *birth canal*)

Parts of the vagina include the:

- Fornix—located at the end of the vagina, its walls curve around the cervix to form a shallow pocket

- Vaginal orifice—the vagina's opening to the outside

- Hymen—a thin barrier of mucous membrane that can extend across the opening of the vagina; it tends to bleed when penetrated or ruptured because it contains blood vessels

5. Female external genitalia—The structures are collectively known as the *vulva*. It consists of the:

- Mons pubis—the outer, elevated, rounded part of the vulva

- Labia majora—two narrow, outer folds of skin

- Labia minora—two inner, hair-free folds of skin that form the outer margins of an area called the *vestibule,* which represents the inner part of the vulva. From anterior to posterior, they contain the:

    - Clitoris—a small projection where the two labia minora meet anteriorly

    - Urinary meatus

- Vaginal orifice—location of a pair of glands, known as *Bartholin's glands,* that provide lubrication to the vagina

- Perineum—the region between the upper border of the vestibule and the anus

6. The mammary glands—organs that produce milk for infant nourishment. Located in the breast, they consist of tissue that is modified from sweat glands. In adult females, the mammary glands undergo enlargement during puberty when estrogens increase in production, directing fat tissue to accumulate between the skin and muscle layer. Portions of the breast of both males and females are:

- The areola—the external, heavily pigmented portion that contains the *nipple*

- 15 to 20 lobes that radiate around the nipple

- The alveolar glands—located in small chambers of each lobe of the gland that produces milk when a woman is lactating, which is under hormonal control

Combining Form	Definition
amni/o, amnion/o	*amnion, amniotic fluid*
andr/o	*male*
balan/o	*glans penis*
cervic/o	*neck, cervix*
chori/o	*membrane, chorion*
cyes/o, cyesi/o	*pregnancy*
embry/o	*embryo*

epididym/o	*epididymis*
episi/o	*vulva*
fet/o	*fetus*
gravid/o, gravidar/o	*pregnancy*
mamm/o, mast/o	*breast*
men/o, menstru/o	*month, menstruation*
pen/o	*penis*
prostat/o	*prostate gland*
semin/o	*seed, sperm*
sperm/o, spermat/o	*seed, sperm*
test/o, testicul/o	*testis or testicle*
urethr/o	*urethra*
vas/o	*vessel, duct*

Factoid

Young females mature as early as age 8. According to one study, by age 8, 48.3% of African-American girls have started developing secondary sexual characteristics. This is true of only 14.7% of white American girls.

INSTRUCTIONAL GOAL: BREAK DOWN AND DEFINE COMMON MEDICAL TERMS USED FOR SYMPTOMS, DISEASES, DISORDERS, PROCEDURES, TREATMENTS, AND DEVICES FOR THE REPRODUCTIVE SYSTEM.

Content Abstract

Signs and Symptoms—Male Reproduction

Prefix	Definition	Combining Form	Definition	Suffix	Definition
a-	*without or absence of*	balan/o	*glans penis*	-ia	*condition of*
oligo-	*little*	orch/o, orchid/o, test/o	*testis*	-itis	*inflammation*
		prostat/o	*prostate gland*	-rrhea	*excessive discharge*
		sperm/o	*sperm, living seed*	-algia	*conditin of pain*
		urethr/o	*urethra*		
		zo/o	*animal, living*		

Medical Term	Definition
Aspermia	the inability to produce or ejaculate sperm, as a symptom of male infertility
Azoospermia	the absence of living sperm in semen, which is a sign of male infertility
Balanorrhea	excessive discharge from the glans penis
Chancres	small ulcers on the skin, a symptom of the sexually transmitted disease *syphilis*
Oligospermia	an abnormally low sperm count, as a sign of male infertility
Papillomas	wartlike lesions on the skin and mucous membranes; a sign of sexually transmitted *human papillomavirus,* commonly known as genital warts
Prostatitis	inflammation of the prostate gland
Prostatorrhea	abnormal, excessive discharge from the prostate gland
Testalgia	testicular pain
Urethritis	inflammation of the urethra, a symptom of the sexually transmitted diseases *Chlamydia* and *trichomoniasis*

Signs and Symptoms—Female Reproduction

Prefix	Definition	Combining Form	Definition	Suffix	Definition
a-	*without or absence of*	colp/o	*vagina*	-algia	*pain*
dys-	*bad, abnormal. painful or difficult*	hemat/o	*blood*	-dynia	*pain*
oligo-	*little*	hydr/o	*water*	-rrhagia	*condition of profuse bleeding or hemorrhage*
poly-	*many*	leuk/o	*white*	-rrhea	*excessive discharge*
		mamm/o, mast/o	*breast*	-salpinx	*trumpet, tube, fallopian tube*
		men/o	*month, menstruation*		
		metr/o	*uterus*		
		py/o	*pus*		

salping/o   *trumpet,*

                   tubefallopia

                   n tube

Medical Term	Definition
Amenorrhea	the absence of menstrual discharge; also called *menostasis*
Colpodynia	symptom of vaginal pain
Colporrhagia	excessive vaginal bleeding
Dysmenorrhea	pain during menstruation
Hematosalpinx	blood in a fallopian tube
Hydrosalpinx	water accumulation in a fallopian tube
Leukorrhea	white or yellowish discharge from the uterus
Mastalgia	pain in the breast
Menometrorrhagia	irregular or excessive bleeding other than during menstruation
Menorrhagia	excessive bleeding during menstruation
Metrorrhagia	bleeding from the uterus at any time other than during normal menstruation
Metrorrhea	discharge of mucus or pus from the uterus

Mittelschmerz		occurs when bleeding from ovulation irritates the peritoneum			
Oligomenorrhea		abnormally reduced discharge during menstruation			
Pyosalpinx		pus within a fallopian tube			

Diseases and Disorders—Male Reproduction

Prefix	Definition	Combining Form	Definition	Suffix	Definition
an-	*without or absence of*	andr/o	*male*	-cele	*hernia, swelling or protrusion*
crypt-	*hidden*	balan/o	*glans penis*	-ism	*condition*
hyper-	*excessive, abnormally high, or above*	epididym/o	*epididymis*	-it is	*inflammation*
		hydr/o	*water*	-pathy	*disease*
		orch/o, orchid/o	*testis*		
		prostat/o	*prostate gland*		

varic/o   *dilated vein*

Medical Term	Definition
Andropathy	a disease that afflicts only males
Anorchism	the absence of one or both testes
Balanitis	inflammation of the glans penis
Benign prostatic hyperplasia	nonmalignant, excessive growth of the prostate gland resulting in constriction of the urethra; symptoms include nocturia, urinary retention, and frequent need to void; also called benign prostatic hypertrophy
Cryptorchidism	the condition of an undescended testis; also called *cryptorchism*
Epididymitis	inflammation of the epididymis
Erectile dysfunction	the inability to achieve or maintain an erection sufficient to perform sexual intercourse, also called *impotency*
Hydrocele	swelling of the scrotum caused by fluid accumulation
Peyronie's disease	induration of the erectile tissue within the

penis, which can cause erectile dysfunction and results in a curvature of the penis if induration is asymmetric

Phimosis a congenital narrowing of the prepuce opening that prevents it from being drawn back over the glans penis; it can be corrected with circumcision; when the glans penis becomes strangulated and produces an emergency situation, the condition is often termed *paraphimosis*

Priapism an abnormally persistent erection of the penis, often accompanied by pain and tenderness, usually caused by drug overdose

Prostate cancer cancer of the prostate gland, which is often highly invasive and metastatic

Testicular torsion a condition of a twisted spermatic cord that causes reduced blood flow to the testis; the affected testis can be lost

Testicular carcinoma cancer originating from a testis, it occurs most often among the 20 to 29-year-old age group; the most common form is called *seminoma,* which arises from spermatogenic cells and metastasizes to nearby lymph nodes

Varicocele   abnormal dilation of the veins of the spermatic cord, caused by failure of the valves within the veins

Diseases and Disorders—Female Reproduction

Prefix	Definition	Combining Form	Definition	Suffix	Definition
a-	*without or absence of*	cerv/o	*cervix*	-al, -ic	*pertaining to*
endo-	*within*	cyst/o	*bladder or sac*	-atresia	*closure or absence of a normal body opening*
ex-	*outside or away from*	lei/o	*smooth*	-cele	*hernia, swelling, or protrusion*
poly-	*excessive, over, or many*	mamm/o, mast/o	*breast*	-ia	*condition of*
		metr/i, hyster/o	*uterus*	-itis	*inflammation*
		my/o	*muscle*	-oma	*tumor*

234

oophor/o, ovar/o	*ovary*	-ptosis	*drop down*
rect/o	*rectum*	-osis	*condition of*
salping/o	*trumpet, tube, fallopian tube*	-pathy	*disease*
vagin/o, colp/o	*vagina*	-s	*plural*
vulv/o	*vulva*		

Medical Term	Definition
Amastia	absence of a breast
Breast cancer	a malignant tumor arising from breast tissue; the most common form is called *infiltrating ductal carcinoma*
Carcinoma in situ	the precancerous form of cervical cancer
Cervical cancer	a malignant tumor arising from the cervix; the most common form is *squamous cell carcinoma*
Cervicitis	inflammation of the cervix
Cystocele	a protrusion of the urinary bladder against the wall of the vagina when the attachments between the two organs weaken
Endocervicitis	inflammation of the inner lining of the cervix
Endometrial cancer	A malignant tumor arising from endometrial tissue of the uterus, it is one form of uterine cancer.

Endometriosis	an abnormal condition of the endometrium (of the uterus), in which endometrial tissue grows in various locations in the pelvic cavity, including on the fallopian tubes, uterus, and ovaries
Endometritis	inflammation of the uterine endometrium
Fibrocystic breast	formation of one or more benign cysts in the breast
Fibroid tumor	a benign tumor containing tissue that arises from the myometrium of the uterus; commonly referred to as *fibroids,* the condition is also called *myoma of the uterus* and *leiomyoma*
Fistula	abnormal passage from one hollow organ to another; a *rectovaginal fistula* occurs between the vagina and rectum, and a *vesicovaginal fistula* occurs between the bladder and vagina
Hysteratresia	closure of the uterus, resulting in an abnormal obstruction in the uterine canal
Mastitis	inflammation of the breast
Mastoptosis	a condition of a sagging breast
Oophoritis	inflammation of an ovary
Ovarian cancer	a malignant tumor arising from an ovary
Parovarian cyst	a cyst of a fallopian tube
Pelvic inflammatory disease	inflammation of the female organs within the pelvic cavity, generally including the ovaries, uterus, and fallopian tubes,

usually caused by bacteria

Premenstrual syndrome	a collection of symptoms, including tension, irritability, mastalgia, edema, and headache, which usually strike during the ten days preceding menstruation
Prolapsed uterus	displacement of the uterus resulting in a downward location, often crowding into the vagina; also called *hysteroptosis*
Salpingitis	inflammation of the fallopian tube
Salpingocele	hernia of the fallopian tube
Toxic shock syndrome	an infectious disease characterized by a rapid onset of symptoms including high fever, skin rash, diarrhea, vomiting, and myalgia, followed by hypotension, leading to shock and, in severe cases, death; it has been linked to noncotton tampon use and is usually caused by S*taphylococcus aureus*
Vaginitis	inflammation of the vagina, which is also known as *colpitis*; in a common form known as *atrophic vaginitis,* the usual symptoms of redness and swelling are accompanied by thinning of the vaginal wall and loss of moisture, usually due to a depletion of estrogens
Vulvitis	inflammation of the external genitals or vulva
Vulvovaginitis	inflammation of the vulva and vagina

Treatments, Procedures, and Devices—Male Reproduction

Prefix	Definition	Combining Form	Definition	Suffix	Definition
anti-	*against*	balan/o	*glans penis*	-al	*pertaining to*
poly-	*excessive, over, or many*	cyst/o	*bladder*	-cele	*protrusion*
trans-	*through, across, or beyond*	hydr/o	*water*	-ectomy	*surgical removal*
		orch/i, orchid/o	*testis*	-pexy	*surgical fixation*
		prostat/o	*prostate gland*	-plasty	*surgical repair*
		ur/o	*urine*	-stomy	*surgical opening*
		vas/o	*vessel*	-tomy	*incision*
		vesicul/o	*small bag*		

Medical Term	Definition
Anti-impotence therapy	collection of therapies to address erectile dysfunction

Balanoplasty   surgical repair of the glans penis

Circumcision   surgical removal of the prepuce, which

                                 involves making a circumscribed incision

                                 around the base of the prepuce

Digital rectal examination   a physical examination that involves the

                                 insertion of a finger into the rectum to feel the

                                 size and shape of the prostate gland through the

                                 wall of the rectum; used to screen the patient

                                 for BPH and prostate cancer

Hydrocelectomy   excision of a hydrocele

Orchidectomy   excision of a testis; also called *orchiectomy;* a

                                 bilateral orchidectomy is commonly called

                                 castration

Orchidopexy   surgical fixation of a testis, which draws an

                                 undescended testis into the scrotum; also called

                                 orchiopexy

Orchidotomy   incision into a testis; also called *orchiotomy*

Orchioplasty   surgical repair of a testis

Penile implant   surgical implantation of a penile prosthesis to

                                 correct erectile dysfunction; available options

                                 include insertion of semi-rigid rods and

	insertion of inflatable balloonlike cylinders
Prostatectomy	excision of a prostate gland to treat cancer and benign prostatic hyperplasia; during a suprapubic prostatectomy, the prostate gland is removed through an incision made above the pubic bone
Prostate-specific antigen	a chemical test that measures blood levels of the protein prostate-specific antigen; elevated levels suggest the probable levels of prostate cancer, so the test is often used to evaluate cancer treatment progress
Prostatocystotomy	incision into the prostate gland and urinary bladder
Transurethral resection of the prostate gland	a surgery to treat BPH when the urethra is obstructed; a more complete surgery than TUIP, it involves the resection of prostate tissue using a retroscope, which is inserted through the urethra; the capsule of the prostate and as much tissue as possible are left intact
Vasectomy	partial excision of the vas deferens, which causes male sterilization; often shortened to vas
Vasovasostomy	a surgery to restore fertility; it involves creating

artificial openings and reconnecting the ends of

the vas where they were severed in an earlier

vasectomy

Vesiculectomy   excision of the seminal vesicles

Treatments, Procedures, and Devices—Female Reproduction

Prefix	Definition	Combining Form	Definition	Suffix	Definition
endo-	*within*	cerv/o	*cervix*	-al, -ic	*pertaining to*
trans-	*through, across, or beyond*	colp/o	*vagina*	-ectomy	*surgical removal*
		episi/o	*vulva*	-gram	*a record or image*
		gynec/o, gyn/o	*woman*	-graphy	*recording process*
		lapar/o	*abdomen*	-logist	*one who studies*
		mamm/o, mast/o	*breast*	-logy	*study of*
		metr/o, hyster/o	*uterus*	-pexy	*surgical fixation*

oophor/o	*ovary*	-plasty	*surgical repair*
path/o	*disease*	-rrhaphy	*suturing*
salping/o	*trumpet, tubefallopian tube*	-scopy	*viewing*
son/o	*sound*	-stomy	*surgical creation of an opening*
vagin/o	*vagina*	-tomy	*incision or to cut*
vulv/o	*vulva*		

Medical Term	Definition
Biopsy	removal of a tissue sample for microscopic evaluation, which may be done by aspiration biopsy, endoscopic biopsy, excisional biopsy, or needle biopsy
Cervical conization	removal of the cone-shaped portion of the cervix
Cervicectomy	excision of the cervix; also called *trachelectomy*
Colpoplasty	surgical repair of the vagina

Colporrhaphy	suture of the vagina
Colposcopy	endoscopic examination of the vagina (and cervix) using a modified endoscope called a *colposcope*
Culdocentesis	surgical puncture into the pelvic cavity to remove fluid from the rectouterine pouch (Douglas' cul-de-sac)
Culdoscopy	endoscopic examination of the space between the rectum and the uterus, called Douglas' cul-de-sac, using a modified endoscope, called a *culdoscope*
Dilation and curettage (D&C)	dilation of the cervix and scraping of the endometrium, in order to control bleeding, obtain a tissue sample for biopsy, or remove polyps
Endometrial ablation	the use of lasers, electricity, or heat to destroy the endometrium, followed by its removal; used to treat abnormal bleeding
Gynecology	the study of diseases of women
Gynecologist	a physician specializing in women's diseases
Hormone replacement therapy	a clinical treatment that includes the replacement of naturally produced hormones with synthetic hormones as a treatment for menopause; also called *estrogen replacement therapy*

Hysterectomy	excision of the uterus, which can include surrounding structures; also called *uterectomy*
Hysteropexy	surgical fixation of the uterus
Hysteroscopy	endoscopic examination of the uterine cavity using a modified endoscope called a *hysteroscope*
Laparoscopy	endoscopic examination of the abdominal or pelvic cavity with a modified endoscope called a *laparoscope*
Mammography	an X-ray procedure that produces an X-ray image of the breast, called a *mammogram*
Mammoplasty	surgical repair of the breasts, resulting in the enlargement or reduction of breast size or removal of a tumor
Mastectomy	excision of a breast; in a simple mastectomy, one entire breast is removed, leaving underlying muscles and lymph nodes intact; a radical mastectomy (also called a *Halstead mastectomy*) is the removal of the entire affected breast, the underlying chest muscles, and local lymph nodes; a modified radical mastectomy is removal of the affected breast and lymph nodes of the underarm,

leaving muscle intact; a lumpectomy is the removal of the cancerous lesions only, in order to conserve the breast

Oophorectomy   excision of an ovary

Panhysterectomy   excision of the uterus, ovaries, and fallopian tubes, which are removed through an abdominal incision; a radical hysterectomy is a similar procedure, in which the lymph nodes, upper portion of the vagina, and surrounding tissue are also removed

Papanicolaou test   a diagnostic procedure in which a sample of cells from the cervix and vagina are removed and examined microscopically for abnormalities; mainly used to screen for cervical cancer; also called *Pap smear* or *Pap test*

Salpingectomy   excision of a fallopian tube

Salpingo-oophorectomy   excision of a fallopian tube and an ovary, usually from the same side

Salpingostomy   surgical creation of an opening through the wall of a fallopian tube, often to treat ectopic tubal pregnancy

Sonohysterography	an ultrasound procedure that records an image of the uterus with the use of sound waves, called a *sonohysterogram*, used to evaluate the postoperative status of polyps, myomas, and adhesions of the uterus
Trachelorrhaphy	suture of the wall of the cervix
Transvaginal sonography	ultrasound procedure in which a probe is inserted into the vagina to record images of the uterus, ovaries, fallopian tubes, and surrounding structures, performed to diagnose ovarian tumors or cysts, monitor pregnancy, and monitor ovulation for treating infertility
Tubal ligation	sterilization procedure by ligating the fallopian tubes
Vaginal speculum	instrument for opening the vaginal orifice to permit visual examination of the vagina and beyond
Vulvectomy	excision of the vulva

Factoids

- It takes about 48 days from the time cells enter meiosis until morphologically mature spermatozoa are formed. Depending on the length of reproduction of spermatogonia (which is not precisely determined), it takes approximately 64 days to complete spermatogenesis.

- Gabriello Fallopio, the Italian anatomist for whom oviducts are named, is also credited for describing and inventing a linen sheath made to fit the glans of the penis. This linen "condom" was devised essentially as a protection against venereal disease.

INSTRUCTIONAL GOAL: BUILD MEDICAL TERMS FROM WORD PARTS ASSOCIATED WITH THE REPRODUCTIVE SYSTEM.

Factoids

- Be careful pronouncing the word *prostate,* which is the male gland surrounding the urethra just below the bladder. The word *prostate* means "one who stands before." The prostate gland "stands before" the exit of urine from the bladder. Many people confuse *prostate* with *prostrate,* which means "to lay flat or in a humbling position." A medical professional must use these two similar-sounding terms correctly.

- Unilateral amastia is often associated with absence of the pectoral muscles. Bilateral amastia is associated in 40% of cases, with multiple congenital anomalies involving other parts of the body as well.

Factoid

Aside from AIDS, the most common and serious complication of sexually transmitted diseases (STDs) among women is pelvic inflammatory disease (PID), an infection of the upper genital tract.

INSTRUCTIONAL GOAL: DEFINE AND SPELL THE WORD PARTS USED TO CREATE TERMS FOR OBSTETRICS.

Abbreviations of the Reproductive Systems, Obstetrics, and STDs

Abbreviation	Definition
AIDS	acquired immunodeficiency syndrome
BPH	benign prostatic hyperplasia
Bx	biopsy
CIN	cervical intraepithelial neoplasia
CIS	carcinoma in situ
C-section	cesarean section
D&C	dilation and curettage
DRE	digital rectal exam
ED	erectile dysfunction
FAS	fetal alcohol syndrome
FBD	fibrocystic breast disease

GYN	gynecology
HBV	hepatitis B virus
HIV	human immunodeficiency virus
HMD	hyaline membrane disease
HPV	human papillomavirus
HRT	hormone replacement therapy
HSV-2	herpes simplex virus type 2
IDC	infiltrating ductal carcinoma
OB	obstetrics
OB/GYN	obstetrics/gynecology
Pap smear (test)	Papanicolaou smear (or test)
PID	pelvic inflammatory disease
PIH	pregnancy-induced hypertension
PMS	premenstrual syndrome
PSA	prostate-specific antigen
RDS	respiratory distress syndrome (of the newborn)
SAB	spontaneous abortion
STI	sexually transmitted infection

TAB	therapeutic abortion
TSS	toxic shock syndrome
TURP	transurethral resection of the prostate
TVS	transvaginal sonography

Factoid

During the embryo's 18th day, the heart starts to beat even as the circulatory system is developing.

INSTRUCTIONAL GOAL: BREAK DOWN AND DEFINE COMMON MEDICAL TERMS USED FOR SYMPTOMS, DISEASES, DISORDERS, PROCEDURES, TREATMENTS, AND DEVICES ASSOCIATED WITH OBSTETRICS.

Content Abstract

Signs and Symptoms—Obstetrics

Prefix	Definition	Combining Form	Definition	Suffix	Definition
dys-	*bad, abnormal, painful, or*	amni/o	*amnion*	-emesis	*vomiting*

	difficult				
hyper-	*excessive, abnormally high, or above*	gravid/o, cyes/o	*pregnancy*	-cyesis	*pregnancy*
poly-	*excessive, over, or many*	lact/o	*milk*	-ia	*pertaining to*
		hydr/o	*water*	-rrhea	*excessive discharge*
		pseud/o	*false*	-s	*plural*
		toc/o	*birth*		

Medical Term	**Definition**
Amniorrhea	abnormal discharge of amniotic fluid
Dystocia	difficult labor
Hyperemesis gravidarum	severe nausea and emesis during pregnancy that can cause severe dehydration in the mother and fetus
Lactorrhea	an abnormal, spontaneous discharge of milk
Polyhydramnios	excessive amniotic fluid

Pseudocyesis   false pregnancy

Diseases and Disorders—Obstetrics

Combining Form	Definition	Suffix	Definition
amni/o, amnion/o	*amnion*	-al	*pertaining to*
blast/o	*germ, bud*	-itis	*inflammation*
chori/o	*chorion*	-osis	*condition of*
erythr/o	*red*	-rrhexis	*rupture*
fet/o	*fetus*	-sis	*state of*
plasm/o	*form*		
tox/o	*poison*		

Medical Term	Definition
Abruptio placentae	premature separation of the placenta from the uterine wall, resulting in either premature birth or fetal death
Amnionitis	inflammation of the amnion
Breech presentation	abnormal childbirth in which the buttocks, feet, or knees emerge first

Cervical effacement	progressive obliteration of the cervical canal during labor
Cleft palate	a congenital abnormality in which the roof of the mouth fails to close during prenatal development, leaving a fissure
Congenital anomaly	an abnormality present at birth
Down syndrome	a congenital disorder caused by a genetic defect in chromosome #21, resulting in degrees of mental retardation and other physical defects; also called *trisomy 21*, and formerly known as *mongolism*
Eclampsia	a condition characterized by convulsions and possible coma, which can follow pre-eclampsia
Ectopic pregnancy	a pregnancy occurring outside the uterus
Embryotocia	expulsion, or abortion, of the embryo
Erythroblastosis	the presence of erthroblasts in the blood. *Erythroblastosis fetalis* is a condition of newborns in which red blood cells are destroyed (by hemolysis) due to an incompatibility between the mother's blood

and the baby's blood occurring when the mother is Rh-negative and has previously had an Rh-positive child, and making subsequent Rh-positive children susceptible.

Fetal alcohol syndrome — a condition caused by alcohol ingestion during pregnancy, it can cause brain dysfunction and growth abnormalities in the newborn

Hyaline membrane disease — a disease of newborns, particularly of premature infants, in which certain cells of the lungs fail to mature by birth, leading to a tendency for lung collapse that can be followed by death; it is also called *respiratory distress syndrome of newborns*

Placenta previa — abnormal attachment of the placenta, in which it is implanted in the lower segment of the uterus near, or obstructing, the external os

Pre-eclampsia — an abnormal development of high blood pressure that can be accompanied by proteinuria and edema, all due to toxemia during pregnancy; also known as *pregnancy-induced hypertension*

Spina bifida

a congenital defect of the vertebral column resulting from an absence of the vertebral arches and often leading to a severe inflammation of the spinal meninges, called a *hydrocele spinalis*

Toxoplasmosis

caused by the bacterium *Toxoplasma gondii*, which can be contracted by exposure to animal feces (most commonly from household pets, such as cats), this disease can cross the placental blood barrier to infect the fetus, causing birth defects or miscarriage

Treatments, Procedures, and Devices—Obstetrics

Combining Form	Definition	Suffix	Definition
abort/o	*miscarry*	-centesis	*surgical puncture*
amni/o	*amnion*	-ic	*pertaining to*
episi/o	*vulva*	-ician	*one who practices*
fet/o	*fetus*	-metry	*measurement or process of measuring*
obstetr/o	*midwife*	-tomy	*incision or to cut*

Medical Term	Definition
Abortion	termination of pregnancy by expulsion of the embryo or fetus from the uterus; a natural expulsion is also called a *miscarriage* or *spontaneous abortion,* and an abortion induced by surgery or drugs is called a *therapeutic abortion*
Amniocentesis	a procedure that involves penetration of the amnion with a syringe and aspiration of a small amount of fluid
Amniorrhexis	rupture of the amnion
Cesarean section	surgical delivery by making an incision through the abdomen and uterus
Contraception	the use of devices and drugs to prevent fertilization, implantation of a fertilized egg or both.
Episiotomy	incision through the perineum, often during labor to prevent its tearing
Fetometry	measurement of the size of the fetus
Obstetrical sonography	ultrasound imaging of the pregnant uterus to

observe fetal development

Obstetrician

a physician practicing in the field of obstetrics

Obstetrics

the medical discipline concerned with

prenatal development, pregnancy, childbirth,

and the 42-day period immediately following

childbirth

Therapeutic abortion contractions

termination of pregnancy by artificially

induced expulsion of the embryo/fetus from

the uterus by means of either surgery or drugs

Sexually Transmitted Infections (STIs)

Medical Term   **Definition**

Acquired immunodeficiency syndrome

acquired mainly through the exchange of

body fluids during sex. Results from the

infection with the human immunodeficiency

virus

Candidiasis

infection by the yeastlike fungus *Candida*

albicans

Chlamydia

its symptoms include urethral or vaginal

discharge and pelvic pain among women,

urethritis and proctitis in men, and

	inflammation of the eye's conjunctiva in newborns that can lead to blindness
Genital herpes	caused by the herpes simplex virus Type 2, or **HSV-2**; it is characterized by periodic outbreaks of ulcerlike sores on the genital and anorectal skin and mucous membranes
Gonorrhea	produces ulcerlike lesions on the mucous membranes and skin of the genital region and is characterized by urethral discharge.
Hepatitis B	a virus that is primarily transmitted via blood exchange, often through blood transfusions or sharing IV needles; it may also be acquired through sexual exchange of body fluids. Hep B causes liver damage that can lead to liver failure and death.
Human papillomavirus	the symptom of papillomas or genital warts, which are transient vesicles on the penis or within the vagina.
Syphilis	caused by a bacterium called a spirochete (*Treponaema pallidum*); it is transmitted by sexual contact and usually first appears as red, painless pustules on the skin that erode to

	form small ulcers known as chancres
Trichomoniasis	an STI caused by the protozoan *Trichomonas*, which is an amoebalike single-celled organism

Factoids

- Cleft palate, along with cleft lip, comprises the fourth most common birth defect in the United States. One of every 700 newborns is affected by cleft lip and/or cleft palate.

- An individual with fetal alcohol syndrome can incur a lifetime health cost of over $800,000.

INSTRUCTIONAL GOAL: BUILD MEDICAL TERMS FROM THE WORD PARTS ASSOCIATED WITH OBSTETRICS AND HUMAN DEVELOPMENT.

Content Abstract

Human development is a continuous process of body change that begins at the moment of fertilization and continues until death. It is divided into two periods as follows:

1. Prenatal development and pregnancy—refers to the changes that occur during the mother's pregnancy, prior to birth:

    - Obstetrics—the clinical field that focuses on this period of life-providing support of the mother throughout pregnancy, during childbirth, and during the first month

or so following childbirth

A new life begins when a gamete from a male and a gamete from a female unite to form a

fertilized egg, or *zygote*. This process is known as *fertilization,* or *conception.* Soon

after single-sperm penetration occurs, the sperm cell's DNA unites with the ovulated

ovum's DNA to form a cell with a complete complement of genetic material. This

marks the beginning of *pregnancy,* or *cyesis,* which normally last 280 days, or 9

calendar months. Pregnancy is divided into three segments, or *trimesters.*

Terms associated with prenatal development include:

- Embryo—Following fertilization, the zygote begins to divide. This term is used to

  denote the cluster of cells and their supporting cells after about two days of

  dividing. The structures include the:

  - Chorion—an outer membrane that eventually unites with the endometrium to

    form the *placenta*

  - Yolk sac—this will later help to form the *umbilical cord*

After about eight days following fertilization, the embryo implants within the

endometrium of the uterus. It increases in size and complexity, and eight weeks later

becomes known as the fetus.

- Fetus—the formed embryo. The fetus continues to grow and develop until its

  weight signals the mother's brain to begin producing the hormone *oxytocin,* which

  stimulates contractions of the uterine wall.

- Labor—once contractions increase in frequency and strength, this process begins.

Labor is a four-stage process as follows:

- Gradual dilation of the cervix

- Increasing strength and duration of uterine contraction

- Birth or *parturition* of the child

- Release of the placenta and other afterbirth materials

- Puerperium—Lasts from the time of delivery of the child until the mother's reproductive organs return to normal.

Factoids

- Abruption occurs in one in about 100 pregnancies, most often in the third trimester, but it can begin any time after 20 weeks of pregnancy. The main symptom is vaginal bleeding, sometimes with uterine discomfort and tenderness, and sudden, continual abdominal pain.

- Thalidomide, used in the 1950s and 1960s as a sedative, led to birth defects in newborns of mothers who took the drug. It has been shown to have clinical activity against Kaposi's sarcoma (KS), an AIDS-related cancer.

Factoid

In 1997 the Institute of Medicine characterized sexually transmitted disease (STDs) as "hidden epidemics of tremendous health and economic consequence" in the United States and recommended an urgent national preventive response. STDs can cause many harmful and often irreversible reproductive problems. STDs can be prevented through behavior modification.

CHAPTER TWELVE

Reproductive System and Obstetrics

Worksheet 1

Phonetic Spelling Challenge

Spell the medical term correctly in the space provided.

1. PRY ah pizm _____

2. pross tah TEK toh mee _____

3. sal pin JYE tiss _____

4. SOO doh sigh EE siss _____

5. SIFF ih liss _____

6. SAL pin JEK toh mee _____

7. EHN doh mee tree OH siss _____

8. mass TEK toh mee _____

9. HIGH droh see LEK toh mee _____

10. SISS toh seel _____

Spelling Challenge: These terms are spelled incorrectly. Spell each term correctly in the space provided.

1. Hydrosalphinx _____

2. Colpectomey _____

3. Vulvaitis   _____

4. Circumsision   _____

5. Epidydimitis   _____

6. Salpingoplexy   _____

7. Fistulla   _____

8. Lapraroscopy   _____

9. Obstretricks   _____

10. Proestrate   _____

Abbreviation Matchup: Select and match the correct abbreviation to the definition.

_____   **1.** pelvic inflammatory disease   **a.** PSA

_____   **2.** cervical   intraepithelial   **b.** BPH

                                  neoplasia

_____   **3.** fibrocystic breast disease   **c.** ED

_____   **4.** prostate-specific antigen   **d.** D&C

_____   **5.** acquired   immunodeficiency   **e.** FBD

                                  syndrome

_____   **6.** digital rectal exam   **f.** TAB

_____   **7.** erectile dysfunction   **g.** AIDS

_____   **8.** benign prostatic hyperplasia   **h.** PID

_____ **9.** therapeutic abortion   **i.** DRE

_____ **10.** dilation and curettage   **j.** CIN

True/False: Mark each statement as true (T) or false (F).

_____ **1.** The reproductive systems of the male and female are subject to infections, tumors, injury, endocrine disorders, and inherited diseases.

_____ **2.** The symptom of testicular pain is known as urethritis.

_____ **3.** A noninvasive diagnostic technique that uses a modified endoscope, called a hysteroscope, to evaluate the uterine cavity is called a laparoscopy.

_____ **4.** A male can elect to become sterile, or unable to produce and ejaculate sperm, by undergoing a vasectomy.

_____ **5.** A surgery to reverse a vasectomy is known as a vesiculectomy.

_____ **6.** Metrorrhagia is the loss of blood from the uterus at any time other than during normal menstruation.

_____ **7.** A bilateral orchidectomy is commonly called castration.

_____ **8.** The most common form of cervical cancer is a squamous cell carcinoma, arising from the epithelial cells lining the opening into the uterus.

_____ **9.** The muscular wall of the uterus is the origin of benign tumors.

_____**10.** In the condition amastia, the individual has more than two elevated areas on the chest or abdomen with areola and nipple.

Fill in the Blank: Fill in the blank with the correct medical term from this chapter.

11. A(n) _____ is a cyst on an ovary.

12. If cyst development spreads into the fallopian tube, the condition is called a(n)

    _____.`

13. Removal   of   the   vagina   is   a   surgery   called   a(n)

    _____.

14. Inflammation   of   the   vagina   is   known   as

    _____.

15. A(n) _____ is a protrusion of the rectum through the

    wall of the vagina.

Short Answer: Write the definition for each of the following terms.

21. Prenatal   _____

22. Urethritis   _____

23. Phimosis   _____

24. Hydrocelectomy _____

25. Menorrhagia   _____

CHAPTER THIRTEEN

THE NERVOUS SYSTEM, MENTAL HEALTH, AND SPECIAL SENSES

LEARNING OBJECTIVES

After completing this chapter, students will be able to:

- Define and spell the word parts used to create terms for the nervous systems.

- Identify the major organs of the nervous system.

- Break down and define common medical terms used for symptoms, diseases, disorders, procedures, treatments, and devices associated with the nervous system, mental health, and the special senses of vision and hearing.

- Build medical terms from the word parts associated with the nervous system, mental health, and the special senses of vision and hearing.

- Pronounce and spell common medical terms associated with the nervous system, mental health, and the special senses of vision and hearing.

INSTRUCTIONAL GOAL: DEFINE AND SPELL THE WORD PARTS USED TO CREATE MEDICAL TERMS FOR THE NERVOUS SYSTEM.

Combining Forms	Definition
blephar/o	*eyelid*
cephal/o	*head*
cerebell/o	*little brain*
cerebr/o, encephal/o	*brain*
conjunctiv/o	*conjunctiva*
crani/o	*skull, cranium*
dacry/o	*tear*
gangli/o	*swelling, knot*
ir/o	*iris*
mast/o	*breast*
mening/i, mening/o	*membrane*
myel/o	*spinal cord*
neur/o	*nerve*
ocul, opt/o, ophthalm/o	*eye*

ot/o	*ear*
phren/o, psych/o	*mind*
radic/o, radicul/o	*nerve root*
retin/o	*retina*
rhin/o	*nose*
scler/o	*sclera; thick or hard*
vag/o	*vagus*
ventricul/o	*ventricle*

Factoid

The average human brain weighs three pounds. The largest brains can be twice that size, but size has no relevance to performance.

INSTRUCTIONAL GOAL: IDENTIFY THE MAJOR ORGANS OF THE NERVOUS SYSTEM.

Content Abstract

1. Brain—the primary organ of the nervous system

2. Nerve impulses—electrochemical impulses

3. Central nervous system (CNS)—includes the brain and spinal cord

4. Peripheral nervous system (PNS)—includes the nerves and sensory receptors

Nervous Tissue

1. Neurons—functional cells of the nervous system

    *   Cell body—contains the nucleus and most of the cytoplasm. Its branches include:

        *   Dendrites—many branches that carry impulses toward the cell body

        *   Axon—a single structure that carries impulses away from the cell body

        *   Neuroglial cells—supportive cells

        *   Schwann cells—a covering for many axons, they form a white-colored protective sheath that is mostly fat; also known as the myelin sheath, it enables an axon to extend great distances

    *   Synapses—the tiny gaps between adjacent neurons. Chemicals known as neurotransmitters are released by a neuron when a nerve impulse reaches its distal end. The chemical diffuses across the synapse to contact the adjacent cell.

    *   Meninges—a thick set of membranes located between the soft nervous tissue of the brain and spinal cord and the hard bones of the cranium. It includes three layers:

        *   Dura mater—a tough, fibrous outer layer

        *   Arachnoid—the middle layer

        *   Pia mater—the thin inner layer; the narrow space located between the arachnoid and pia mater is called the subarachnoid space

Central Nervous System—the central station for incoming and outgoing nerve impulses

1. Brain—the organ that receives sensory information, interprets and integrates this information, and controls muscles and glandular response. It consist of the following main parts:

   - Cerebrum—the largest, most significant part of the brain. It is divided into right and left cerebral hemispheres. The fissure that separates the two halves is bridged by a band of nervous tissue called the corpus callosum.

     - Cerebral cortex—an outer fringe of gray matter that is the most important functional part of the cerebrum.

     - Lobes—the functional zones of the cerebral cortex. Each lobe houses a cluster of neurons that perform a particular function in common, such as the interpretation of sound, the control of voluntary muscles, the perception of sensory information from the skin, or the formation of personality traits.

   - Cerebellum (little brain)—a small, convoluted mass located below and posterior to the cerebrum that coordinates muscle responses and manages equilibrium.

   - Diencephalon (double brain)—located beneath the cerebrum and anterior to the cerebellum. Contains the following two areas:

     - Thalamus (inner chamber)—located in the approximate center of the brain and acts as a relay station, redirecting nerve impulses to and from the cerebrum

     - Hypothalamus—located below the thalamus. It is the center for involuntary autonomic activities, such as regulation of heartbeat, thirst, blood pressure, and glandular functions.

- Brain stem—located below the diencephalon. It includes:

  - Medulla—transmits nerve impulses between the spinal cord and brain and regulates breathing rhythms

  - Pons (bridge)—provides a connection between the medulla and cerebellum

- The brain also consists of ventricles (little belly)—small spaces within the brain's center that are filled with cerebrospinal fluid, a slightly yellowish fluid that is continuously produced from the blood supply.

2. Spinal cord—an extension of the medulla, which terminates between L1 and L2. It is divided into 31 segments, each consisting of a pair of spinal roots, which form 31 pairs of spinal nerves. The spinal cord consists of both gray matter (center) and white matter (the outer portion of the cord consisting of long nerves that carry impulses up and down the spinal cord).

Peripheral Nervous System—nerves and sensory receptors

1. Nerves—consist of branches from the 12 cranial nerves that communicate with the brain and branches of the 31 pairs of spinal nerves that communicate with the spinal cord. Each nerve consists of:

- Nervous tissue—Consists of axons of neurons, which are bundled together and wrapped with layers of connective tissue.

- Sensory or afferent nerves—Carry nerve impulses from the sensory receptors to the brain or spinal cord.

- Motor or efferent nerves—Carry impulses from the brain or spinal cord to muscles and glands.

- Ganglia—clusters of neuron cell bodies that lie outside the brain and spinal cord. They are centers where nerve impulses are passed from one neuron to another across synapses.

2. Sensory receptors—nervous structures that respond to changes in the environment.

Special Senses

The special sense of vision begins when the eyes detect light. Light is channeled to specialized sensory receptor cells deep in each eye. Once stimulated, the receptor cells send signals to the brain where the interpretation of light into an image is made. The organs and their parts include:

1. Eyes—the organs of vision; each is set inside the orbit of the skull.

- The eyes are protected by:

  - Eyelids—a protective fold

  - Conjunctiva—a thin sheet of cells that covers the anterior surface of the eye and the inner surface of the eyelid

  - Lacrimal glands—keeps the conjunctiva moist with a watery secretion called tears

  - Meibomian glands—located along the edge of the eyelids and provide moisture to the eye

- Internal eye structures consist of three layers:

  - Fibrous layer—the outermost layer

- Sclera—the white portion of the eye

- Cornea—the transparent window of the eye

- Anterior chamber—a narrow chamber behind the cornea that is filled with a watery fluid called aqueous humor, which is constantly produced and reabsorbed.

- Vascular layer—the middle layer of the eye named because of the abundance of blood vessels that it contains. It includes:

  - Iris—the colored ring of the eye, and includes smooth muscle fibers that regulate the amount of light entering the eye through the pupil.

  - Pupil—the black opening in the center of the eye.

  - Lens—a transparent disk that is suspended behind the pupil by suspensory ligaments, which attach to ciliary muscles. The ciliary muscles pull on the lens when they contract, changing the lens shape to allow light to focus on the retina. Behind the lens is a large cavity, the posterior cavity, which is filled with a gelatinous material known as vitreous humor.

  - Choroid—a blood vessel-rich area whose blood supply nourishes the cells of the retina.

- Nervous layer—the innermost layer of the eye consisting of:

  - Retina—a thin film at the back of the eye that is composed of neurons, including receptors sensitive to light known as photoreceptors.

  - Photoreceptors—includes two types of cells.

- Rods—photoreceptors that are very sensitive to small amounts of light, but are limited to black and white shades.

- Cones—photoreceptors that require more light as stimulus, but enable you to perceive color.

- Fovea centralis—the area of the retina that is the site for your sharpest vision. It contains the highest concentration of cones.

- Optic nerve—the path of retinal information to the brain.

- Optic disk—the exit point of the optic nerve in the retina that lacks photoreceptors. It is called the "blind spot."

2. Ears—the organ of hearing or *audation,* which contain sensory receptors that respond to sound waves, or mechanical vibrations, by generating a nerve impulse.

- Mastoid process—where the inner ear is located

- Outer ear—flaplike appendages on the sides of your head; also called auricles

- External auditory canal—the passage from the external opening to the eardrum

- Cerumen—earwax, secreted by specialized glands

- Middle ear—consists of the tympanic membrane and ossicles

  - Tympanic cavity—air-filled cavity between the inside wall of the eardrum and the inner ear

  - Eustachian tube

  - Malleus or hammer—ossicle

- Incus or anvil—ossicle

- Stapes or stirrup—ossicle

- Labyrinth—channels within the temporal bone. Consists of:

  - Cochlea

  - Vestibule

  - Semicircular canals

  - Organ of Corti—contains sensory receptors

Factoid

The brain has the natural ability to calm itself through the use of substances called endorphins and enkephalins. The brain itself produces these substances in response to stress. The substances are very much like morphine in that they reduce anxiety, reduce pain, and produce a sense of general well-being. We do not understand exactly how these substances work.

INSTRUCTIONAL GOAL: BREAK DOWN AND DEFINE COMMON MEDICAL TERMS USED FOR SYMPTOMS, DISEASES, DISORDERS, PROCEDURES, TREATMENTS, AND DEVICES ASSOCIATED WITH THE NERVOUS SYSTEM.

Signs and Symptoms

Prefix	Definition	Combining Forms	Definition	Suffixes	Definition
a-	*without or absence of*	cephal/o	*head*	-algesia	*condition of pain*
hyper-	*excessive, abnormally high, or above*	esthesi/o	*sensation*	-algia	*pain*
hypo-	*under, abnormally low, or below*	neur/o	*nerve*	-asthenia	*weakness*
par, para-	*alongside or abnormal*	phasi/o	*to speak*	-ia,-a	*condition of*
poly-	*excessive, over, or many*				

Medical Term	Definition
Aphasia	inability to speak
Cephalagia	a headache, or general pain to the head

Convulsion	a series of involuntary muscular spasms caused by an uncoordinated excitation of motor neurons that trigger muscle contraction
Dysphasia	difficulty speaking
Hyperesthesia	increased sensitivity to stimulation such as touch or pain
Hyperalgesia	extreme lack of sensitivity to normally painful stimuli
Neuralgia	pain in a nerve
Neurasthenia	a vague condition of body fatigue often associated with depression
Paresthesia	an abnormal sensation of numbness and tingling without an objective cause
Polyneuralgia	a sysmptom of pain in many nerves
Syncope	a temporary loss of consciousness due to a sudden reduction of blood flow to the brain

Diseases and Disorders

Prefix	Definition	Combining Forms	Definition	Suffix	Definition
a-	*without or absence of*	ather/o	*fatty substance or plaque*	-al,-ar,-ic,-i on,-uss	*pertaining to*
epi-	*upon, above, or top*	aut/o	*self*	-cele	*hernia, swelling, or*

					protrusion
para-	*beside, next to*	cephal/o	*head*	-ism,-osis	*condition of*
poly-	*excessive, over, or many*	cerebell/o	*little brain, cerebellum*	-it is	*inflammation*
		cerebr/o, encephal/o	*brain*	-lepsy	*seizure*
		embol/o	*a plug*	-malacia	*softening*
		gli/o	*glue*	-oma	*tumor*
		gnos/o	*knowledge*	-pathy	*disease*
		later/o	*side*	-plegia	*paralysis*
		mening/i, mening/o	*membrane*	-rrhage	*profuse bleeding or hemorrhage*
		myel/o	*spinal cord*	-troph	*development*
		narc/o	*numbness*		
		neur/o	*nerve*		
		scler/o	*thick, hard, sclera*		
		poli/o	*gray*		
		thromb/o	*clot*		

ventricul/o *ventricle*

Medical Term	Definition
Agnosia	a loss of the ability to interpret sensory information
Alzheimer disease	deterioration of brain function characterized by confusion, short-term memory loss, and restlessness
Amyotrophic lateral sclerosis	progressive atrophy of muscle caused by hardening of nervous tissue on the lateral columns of the spinal cord. It is known as *Lou Gehrig's disease* after the professional baseball player whose experience with this disease brought it to national attention.
Autism	developmental disorder that varies in its severity with the patient; characterized by withdrawal from outward reality and impaired development in social conduct and communication
Bell palsy	a condition of muscular paralysis
Cerebellitis	inflammation of the cerebellum
Cerebral aneurysm	a type of cerebrovascular disease where a blood vessel supplying the brain becomes dilated
Cerebral arteriosclerosis	a type of cerebral vascular disease characterized by hardening of the arteries of the brain
Cerebral embolism	presence of an embolism in a blood vessel supplying the brain

Cerebral palsy | a condition revealed by partial muscle paralysis that is caused by a brain defect or lesion present at birth or shortly after birth

Cerebral thrombosis | a thrombosis (lodged blood clot) within blood vessels supplying the brain

Cerebrovascular accident | caused by a thrombosis, embolism, or hemorrhage; this disruption of the blood supply to the brain results in functional losses or death; also called *stroke*

Coma | a general term describing several levels of decreased consciousness; also known as *deep sleep*

Concussion | an injury to soft tissue resulting from a blow or violent shaking

Encephalitis | inflammation of the brain, usually caused by bacterial or viral infection

Encephalomalacia | softening of brain tissue, usually caused by deficient blood flow

Epilepsy | a brain disorder characterized by recurrent seizures

Gangliitis | inflammation of a ganglion

Glioma | a tumor of neuroglial cells

Hydrocephaly | increased volume of CSF in the brain ventricles of a child before the cranial sutures have sealed, causing enlargement of the cranium

Meningioma | benign tumor of the meninges

Meningitis	inflammation of the meninges, usually caused by bacterial or viral infection
Meningocele	protrusion of the meninges through an opening caused by a defect in the skull or spinal column
Meningomyelocele	protrusion of the meninges and the spinal cord through the spinal column
Multiple sclerosis	the deterioration of the myelin sheath covering axons within the brain, exhibited by episodes of localized functional losses
Myelitis	inflammation of the spinal cord
Narcolepsy	a sleep disorder characterized by sudden uncontrollable attacks of sleep, attacks of paralysis, and hypnagogic hallucinations
Neuritis	inflammation of a nerve
Neuroarthropathy	a disease of the nervous system resulting in pain within one or more joint
Neuroma	a general term for any tumor originating from nervous tissue
Neurosis	an emotional disorder involving a counterproductive way of dealing with stress
Paraplegia	paralysis from the waist down
Parkinson disease	chronic degenerative disease of the brain indicated by hand tremors, rigidity, expressionless face, and shuffling gait; also

called *Parkinsonism*

Poliomyelitis	inflammation of gray matter of the spinal cord caused by one of several polioviruses that often leads to paralysis; also called *polio*
Polyneuritis	inflammation of many nerves at one time
Quadriplegia	paralysis of all four limbs; also known as *tetraplegia*
Rabies	acute, often fatal infection of the central nervous system that is caused by a virus transmitted to humans by the bite of an infected animal
Ventriculitis	condition of inlammation of the ventricles of the brain

Treatments, Procedures, and Devices

Prefix	Definition	Combining Forms	Definition	Suffix	Definition
an-	*without or absence of*	angi/o	*vessel*	-algesia	*pain*
epi-	*upon, over, above, or on top*	cerebr/o	*brain, cerebrum*	-ectomy	*surgical excision or removal*

crani/o	*skull, cranium*	-al,-ic	*pertaining to*
dur/o	*hard*	-gram	*a record or image*
ech/o	*to bounce*	-graphy	*recording process*
electr/o	*electricity*	-iatry	*treatment or specialty*
encephal/o	*brain*	-ist	*one who specializes*
esthesi/o	*sensation*	-logy	*study of*
gangli/o	*swelling or knot*	-lysis	*loosen or dissolve*
myel/o	*spinal cord*	-plasty	*surgical repair*
neur/o	*nerve*	-rrhaphy	*suturing*
psych/o	*mind*	-tome	*cutting instrument*
radic/o, rhiz/o	*nerve root*	-tomy	*incision or to cut*
tom/o	*to cut*		
vag/o	*vagus, nerve*		

Medical Term	Definition
Analgesic	an agent that relieves pain
Anesthesia	without feeling or sensation
Cerebral angiography	X-ray photography of the blood vessels in the brain following injection of a contrast medium
Computed tomography	also called a *CT scan,* this procedure involves the use of a computer to interpret a series of images and construct them from a three-dimensional view of the brain; this is particularly useful in diagnosing tumors
Craniectomy	excision of part of the skull to approach the brain
Craniotomy	incision into the skull to approach the brain
Echoencephalography	use of *ultrasonography* or *ultrasound* to record brain structures
Electroencephalography	a procedure recording the electrical impulses of the brain
Evoked potential studies	also called EP studies, this group of diagnostic tests measures changes in brain waves during particular stimuli to determine brain fuction, providing a test for sight, hearing, and other senses
Epidural	the injection of a spinal block anesthetic into the epidural space external to the spinal cord

Ganglionectomy	excision of a ganglion; also called *gangliectomy*
Lumbar puncture	aspiration of CSF from the subarachnoid space in the lumbar region of the spinal cord
Magnetic resonance imaging	the use of magnets to identify structural details of soft tissue within the body, coupled with computer imaging to produce three-dimensional images that are useful in targeting brain tumors, brain trauma, MS, and other conditions
Myelogram	X-ray photography of the spinal cord following the injection of a contrast dye
Neurectomy	excision of a nerve
Neurology	the study and medical practice of the nervous system
Neurolysis	separating a nerve by removing adhesions
Neuroplasty	surgical repair of a nerve
Neurorrhaphy	suture of a nerve
Neuroscientist	one who studies in the field of neuroscience
Neurotomy	incision into a nerve
Neurologist	a physician specializing in general disorders of the nervous system
Positron emission tomography	a brain scan providing a map of blood flow within the

	brain that can be correlated to brain activity
Psychiatry	the branch of medicine that addresses disorders of the brain
Psychology	the field of study of human behavior
Psychotherapy	technique used in treating behavioral and emotional issues
Radicotomy	incision into a nerve root; also called *rhizotomy*
Reflex testing	a diagnostic test performed to observe the body's response to a touch stimulus, which is useful when assessing stroke, head trauma, and other neurological challenges; this includes *deep tendon reflexes* involving percussion at the patellar tendon or elsewhere, and *Babinski reflex*, involving stimulation of the plantar surface of the foot
Vagotomy	when several branches of the vagus nerve are severed to reduce acid secretion into the stomach in an effort to prevent the reoccurance of peptic ulcer

Factoid

Schwann cells are prime candidates for strategies to bridge areas of spinal cord injury. Stimulation of Schwann cells to divide will be needed to cultivate the large numbers of cells to be used in grafts to promote spinal cord regeneration.

INSTRUCTIONAL GOAL: BUILD MEDICAL TERMS FROM THE WORD PARTS ASSOCIATED WITH THE NERVOUS SYSTEM.

Factoids

- The severity of dementia in Alzheimer's patients is directly related to the reduction of the amount of the neurotransmitter acetylcholine.

- The most common form of ALS in the United States is "sporadic" ALS. It may affect anyone, anywhere. "Familial" ALS (FALS) means the disease is inherited.

- Treatments of epilepsy seek to reduce the frequency of seizures or prevent their occurrence.

INSTRUCTIONAL GOAL: PRONOUNCE AND SPELL COMMON MEDICAL TERMS ASSOCIATED WITH THE NERVOUS SYSTEM.

Factoids

- Meningiomas account for 13% to 17% of intracranial tumors in the United States. Multiple meningiomas are 1% to 6% of this total.

- The sciatic nerve is the largest nerve in the human body, about the diameter of a pencil.

- TIAs occur in approximately 50% to 80% of patients who've had a cerebral infarct from thrombosis.

- Magnetic resonance angiography (MRA) is a noninvasive way to evaluate the arteries and veins throughout the body. This procedure doesn't require threading a catheter into arteries, as does traditional angiography.

INSTRUCTIONAL GOAL: BREAK DOWN AND DEFINE COMMON MEDICAL TERMS USED FOR SYMPTOMS, DISEASES, DISORDERS, PROCEDURES, TREATMENTS, AND DEVICES ASSOCIATED WITH MENTAL HEALTH DISEASE AND DISORDERS, THE EYES, AND THE EARS.

Mental Health Diseases and Disorders

Prefix	Definition	Combining Forms	Definition	Suffix	Definition
bi-	*two*	neur/o	*nerve*	-ia	*condition of*
dys-	*bad, abnormal, painful or difficult*	psych/o	*mind*	-ic	*pertaining to*
		schiz/o	*to divide or split*	-lexia	*pertaining to a word or phrase*
		somat/o	*body*	-mania	*madness or frenzy*
				-pathy	*disease*
				-phobia	*fear*
				-sis	*state of*

Medical Term	Definition
Anxiety disorder	when fear dominates behavior
Attention deficit disorder	a neurological disorder characterized by short attention span and poor concentration
Bipolar disease	causes alternating periods of high energy and mental confusion
Dementia	an impairment of mental function characterized by memory loss, disorientation, and confusion
Dyslexia	a reading handicap in which some letters and numbers are reversed in order by the brain
Mania	an emotional disorder of abnormally high psychomotor activity
Neurosis	an emotional disorder involving a counterproductive way of dealing with mental stress
Paranoia	disorder where a person experiences persistent delusions of persecution resulting in mistrust and combativeness
Phobia	an irrational, obsessive fear
Posttraumatic stress disorder	a severe mental strain or emotional trauma causes sleeplessness, anxiety, and paranoia
Psychopathy	a general term for a mental or emotional disorder
Psychosis	an individual suffering from a gross distortion or disorganization of their mental capacity, emotional response, and capacity to

recognize reality and relate to others may be diagnosed with this disease

Psychosomatic   refers to the influence of the mind over bodily functions, especially disease

Schizophrenia   characterized by delusions, hallucinations, and extensive withdrawal from other people and the outside world.

Eye Diseases and Disorders

Prefix	Definition	Combining Forms	Definition	Suffix	Definition
a-	*without or absence of*	blephar/o	*eyelid*	-iasis	*condition of*
dipl-	*double*	conjunctiv/o	*conjunctiva*	-ism	*condition of*
hyper-	*excessive, abnormally high, or above*	cyst/o	*bladder or sac*	-malacia	*softening*
		dacry/o	*tear*	-opia	*condition of vision*
		ir/o	*iris*	-pathy	*disease*

kerat/o	*hard, cornea*	-plegia	*paralysis*
lith/o	*stone*	-ptosis	*condition of falling or drooping*
my/o	*muscle*	-rrhagia	*condition of profuse bleeding or hemorrhage*
ophthalm/o	*eye*		
presby/o	*old age*		
retin/o	*retina*		
sinus/o	*sinus cavity*		
stigmat/o	*point*		

Medical Term	Definition
Blepharoptosis	drooping of an eyelid
Cataract	a reduction of transparency of the lens
Chalazion	a localized swelling at the edge of an eyelid caused by obstruction of a meibomian gland
Conjunctivitis	inflammation of the conjunctiva

Dacryocystitis	inflammation of a lacrimal sac
Detached retina	separation of the retina from the choroids at the back of the eye
Diplopia	double vision
Glaucoma	a loss of vision resulting from increased intraocular pressure, which damages the optic nerve
Hordeolum	an infection of a meibomian gland causing a local swelling of the eyelid
Hyperopia	reduced vision with nearby objects; also called farsightedness
Iritis	inflammation of the iris
Keratitis	inflammation of the cornea
Macular degeneration	progressive deterioration of an area of the retina known as the macula lutea, leading to a loss of central vision; the most common cause is age, which is called *age-related macular degeneration*
Myopia	reduced vision with distant objects; also called nearsightedness
Nyctalopia	poor vision at night or in dim light
Ophthalmopathy	a general term for a disease of the eye
Ophthalmoplegia	paralysis of the eye affecting the muscles that move the eyeball
Presbyopia	impaired vision due to aging
Retinopathy	a general term for any disease of the retina

Eye Treatments, Procedures, and Devices

Prefix	Definition	Combining Forms	Definition	Suffix	Definition
intra-	*within*	dacry/o	*tear*	-ar	*pertaining to*
		cyst/o	*bladder or sac*	-logist	*one who studies*
		kerat/o	*hard, cornea*	-metrist	*one who measures*
		ocul/o	*eye*	-stomy	*surgical creation of an opening*
		opt/o	*eye*	-tomy	*incision or to cut*
		rad/I	*spoke of a wheel*		
		rhin/o	*nose*		

Medical Term	Definition
Cataract extraction	excision of a lens that has lost its clarity
Corneal grafting	procedure where the injured cornea is removed and replaced by implantation of a donor cornea
Dacryocystorhinostomy	procedure where a channel is surgically created between the nasal cavity and lacrimal sac to promote drainage.
LASIK	acronym for *laser-assisted in situ keratomileusis;* it is the use of a laser to reshape the corneal tissue beneath the surface of the cornea to correct vision abnormalities, such as *myopia, hyperopia,* and *astigmatism*
Ophthalmologist	a physician specializing in the study of eyes
Ophthalmology	the field of medicine focusing on the study of disease related to the eyes
Ophthalmoscopy	use of a hand-held instrument with a light, called an ophthalmoscope to view the eye's interior
Optician	a technician trained in filling prescriptions for corrective lenses
Optometry	measurement of vision, usually to test acuity for prescribing corrective lenses; the process includes the use of an *optometer*, which measures the range and sharpness of vision
Optometrist	a professional—not a physician—trained to examine eyes to

correct vision problems and eye disorders

Factoids: Diplopia is the medical term for double vision. In multiple sclerosis, it is usually caused by lesions in the brain stem where the cranial nerves serving the eye muscles arise. The most common types of infantile nystagmus are congenital nystagmus (CN) and latent/manifest latent nystagmus (LMLN).

Ear Diseases and Disorders

Prefix	Definition	Combining Form	Definition	Suffix	Definition
(none)		extern/o	*exterior*	-a	*singular*
		mast/o	*breast, mastoid*	-algia	*pain*
		med/o	*middle*	-it is	*inflammation*
		ot/o	*ear*	-oid	*resembling*
		scler/o	scler/o*thick or hard, sclera*	-osis	*condition of*
				-pathy	*disease*
				-rrhea	*excessive discharge*
				-scopy	*process of*

Medical Term	Definition
Mastoiditis	inflammation of the mastoid process and the associated tissues
Ménière's disease	a chronic disease of the inner ear that includes symptoms of dizziness and ringing in the ears
Myringitis	inflammation of the eardrum; also called *tympanitis*
Otitis externa	inflammation of the external auditory canal
Otitis media	inflammation of the middle ear
Otopathy	disease of the ear
Otosclerosis	abnormal formation of bone between the stapes and the oval window, causing a progressive loss of hearing
Vertigo	the sensation of whirling motion

Factoid

The organ of Corti is known as the true organ of hearing. It contains some 16,000 to 20,000 hair cells distributed along the basilar membrane, which follows the spiral of the cochlea.

Vertigo is dizziness that creates the sense that you or your surroundings are spinning or moving. It is defined as a false illusion of motion with a distinct sensation of rotation

("The room was spinning around me").

Labyrinthectomy of the affected ear can be quite effective for benign positional vertigo. The handheld tympanometer is a device that provides quantitative information on the function of structures and the presence of fluid in the middle ear. The graphic display of this data is the tympanogram.

Abbreviations.

Abbreviation	Definition
AD	Alzheimer's disease
ADHD	attention deficit hyperactivity disorder
ADD	attention deficit disorder
ALS	amyotrophic lateral sclerosis
Ast	astigmatism
CT (CAT) scan	computed (axial) tomography scan
CP	cerebral palsy
CVA	cerebrovascular accident (stroke)
DTR	deep tendon reflexes
EchoEG	echoencephalography
EEG	electroencephalogram
Em	emmetropia

EP   evoked potential

IOL   intraocular lens

LASIK   laser-assisted in situ keratomileusis

MRI   magnetic resonance imaging

MS   multiple sclerosis

OM   otitis media

PD   Parkinson disease

PET   positron emission tomography

PTSD   posttraumatic stress disorder

CHAPTER THIRTEEN

The Nervous System, Mental Health, and Special Senses

Worksheet 1

Phonetic Spelling Challenge: Spell the medical term correctly in the space provided.

1. pall ee noo ROH path ee   _____

2. RAE beez   _____

3. ek oh en SEFF ah LOG rah fee   _____

4. NOO roh plass tee   _____

5. ceh REE bral anj ee OHG rah fee   _____

6. LUM bar PUNK shur   _____

7. noo ROT oh mee   _____

8. glaw KOH mah   _____

9. LAY sik   _____

10. SKIZ oh FREHN ee ah   _____

Spelling Challenge: These terms are spelled incorrectly. Spell each term correctly in the space provided.

1. Gliomae   _____

2. Hydrocephaelus   _____

3. Incephalitis   _____

4. Oldheimer's _____

5. Singcope _____

6. Afasia _____

7. Cephalalglia _____

8. Parresthesia _____

9. Anneurysm _____

10. Menninjocele _____

Abbreviation Matchup: Select and match the correct abbreviation to the definition.

_____ **1.** astigmatism **a.** ADD

_____ **2.** positron emission tomography **b.** CVA

_____ **3.** cerebral palsy **c.** DTR

_____ **4.** amyotrophic lateral sclerosis **d.** Ast

_____ **5.** Parkinson disease **e.** ADHD

_____ **6.** electroencephalogram **f.** PD

_____ **7.** attention deficit disorder **g.** PET

_____ **8.** deep tendon reflex **h.** ALS

_____ **9.** attention-deficit hyperactivity disorder **i.** CP

_____ **10.** cerebral vascular accident **j.** EEG

True/False: Mark each statement as true (T) or false (F).

_____ **1.** To treat dacryocystitis, antibiotic eye drops are often used to defeat the bacterial infection.

_____ **2.** The condition of double vision is called glaucoma.

_____ **3.** The lens of the eye is normally transparent.

_____ **4.** Correcting vision disorders is usually attempted by the use of corrective lenses or contact lenses following a vision examination by an optometrist.

_____ **5.** Nearsightedness is called hyperopia.

_____ **6.** Emmetropia is the normal condition of the eye.

_____ **7.** Hordeolum is also called a "sty."

_____ **8.** Ophthalmomalacia means hardening of the eye.

_____ **9.** The most common form of psychosis is psychosis.

_____**10.** Acrophobia is the abnormal fear of heights.

Fill in the Blank: Fill in the blank with the correct medical term from this chapter.

11. _____ is an obsessive concern with fire.

12. The term _____ literally means "pertaining to body and mind."

13. In the condition known as _____, transparency of the lens is reduced, usually as a normal part of the aging process.

14. A surgical incision into a nerve root is called

_____.

15. The most common form of pain management is the use of

_____.

Short Answer: Write the definition for each of the following terms.

21. Autism   _____

_____

22. Alzheimer, disease   _____

_____

23. Cerebral atherosclerosis_____

_____

24. Mania   _____

_____

25. Anxiety disorder   _____

_____

CHAPTER FOURTEEN

THE ENDOCRINE SYSTEM

LEARNING OBJECTIVES

After completing this chapter, students will be able to:

- Define and spell the word parts used to create terms for the endocrine system.

- Identify the major organs of the endocrine system and describe their structure and function.

- Break down and define common medical terms used for symptoms, diseases, disorders, procedures, treatments, and devices associated with the endocrine system.

- Build medical terms from the word parts associated with the endocrine system.

- Pronounce and spell common medical terms associated with the endocrine system.

INSTRUCTIONAL GOAL: DEFINE AND SPELL THE WORD PARTS USED TO CREATE MEDICAL TERMS FOR THE ENDOCRINE SYSTEM.

Content Abstract

Combining Form	Definition
aden/o	*gland*
adren/o	*adrenal gland*
crin/o	*to secrete*
gonad/o	*sex gland*
hormon/o	*to set in motion*
pancreat/o	*sweetbread, pancreas*
ren/o	*kidney*
thyr/o, thyroid/o	*shield, thyroid*

Factoid

Note the correct spelling forms of mucus: as a noun, it is spelled m-u-c-u-s, but when used as an adjective, such as "mucous membrane," the spelling changes to m-u-c-o-u-s.

INSTRUCTIONAL GOAL: IDENTIFY THE MAJOR ORGANS OF THE ENDOCRINE SYSTEM AND DESCRIBE THEIR STRUCTURE AND FUNCTION.

Content Abstract

1. **Pituitary gland**—located within the cranial cavity immediately below the brain and connected to the hypothalamus. It is also called the hypophysis. It consists of two lobes:

   - Anterior lobe, or adenohypophysis—made up of soft glandular tissue that produces the following hormones:

     - Growth hormone (GH)—regulates metabolism and body growth

     - Adrenocorticotrophic hormone (ACTH)—activates the adrenal gland

     - Melanocyte-stimulating hormone (MSH)—stimulates skin pigment production

     - Thyroid-stimulating hormone (TSH)—stimulates the thyroid gland

     - Prolactin (PRL)—stimulates milk secretion by the mammary glands

     - Follicle-stimulating hormone (FSH)—stimulates development of ova and sperm

     - Luteinizing hormone (LH)—stimulates secretion of sex hormones by the gonads

   - Posterior lobe, or neurohypophysis—consists of nervous tissue that secretes two hormones:

- Oxytocin (OT)—stimulates contractions of the uterus and milk secretion by the mammary glands

- Antidiuretic hormone (ADH)—stimulates water reabsorption by the kidneys, and reduces urine volume

2. **Pineal gland**—located in the cranial cavity, in the center of the brain. It secretes:

- Melatonin—regulates body rhythms including sleep cycles

3. **Thyroid gland**—Located in the anterior part of the neck, it is a butterfly-shaped endocrine organ that wraps around the larynx. It consists of right and left lobes connected by a narrow band called the isthmus. It secretes the following three hormones:

- Thyroxine (T4)—regulates the breakdown of glucose and the synthesis of most cells in the body

- Triiodothyronine (T3)—works in conjunction with thyroxine to produce the same effects (T4 and T3 are referred to as thyroid hormone)

- Calcitonin (CT)—stimulates the production of new bone material to reduce calcium in the blood

4. **Parathyroid glands**—located in the anterior part of the neck, they consist of three or four pea-sized organs embedded within the posterior side of the larger thyroid gland. They secrete:

- Parathyroid hormone (PTH)—increases calcium levels in the blood, the opposite effect of calcitonin

5. **Adrenal glands**—located within the abdominal cavity on the top of each kidney. They also are called the suprarenal glands. Each gland includes a cortex, or outer part, and a medulla, or inner part. The adrenal cortex produces:

- Aldosterone—regulates body fluid balance and blood pressure

- Glucocorticoids—reduce inflammation

- Androgens and estrogens—stimulate the development of sex characteristics

The adrenal medulla secretes two hormones. They are:

- Epinephrine, or adrenaline—prolongs the condition for the "fight or flight" response, which includes an increase in metabolism, heart rate, blood pressure, etc.

- Norepinephrine—a neurotransmitter that is a precursor to the release of epinephrine; norepinephrine also prolongs the condition for the "fight-or-flight" response, which includes an increase in metabolism, heart rate, blood pressure, etc.

6. **Pancreas**—located within the abdominal cavity, immediately behind the stomach. It is a soft, oblong organ that performs two functions. They are the:

- Secretion of hormones—hormones are secreted by a part of the pancreas consisting of clusters of cells called the islets of Langerhans. These clusters are distributed throughout the organ. Each islet secretes two primary hormones:

  - Insulin—reduces blood sugar levels by stimulating the conversion of glucose to glycogen and facilitating the uptake of glucose into body cells.

- Glucagon—increases blood sugar levels by stimulating the conversion of glycogen into glucose in the liver, releasing glucose into the bloodstream.

- Secretion of digestive enzymes—will be discussed in the chapter on the digestive system.

7. **Thymus gland**—a soft gland that shrinks in size after puberty; it is located anterior and above the heart in the thoracic cavity. It secretes:

- Thymosin—helps to establish the immune response during early childhood

8. **Gonads**—produce sex hormones and the reproductive cells, or gametes. They are discussed in the chapter on reproduction, but are the:

- Testes—the male gonads, which secrete testosterone

- Ovaries—the female gonads, which secrete estrogen and progesterone

- Endocrine cells located in the heart, stomach, and kidneys, to be discussed in other chapters

Factoid

People with an affected parent or sibling are at 3.5 times greater risk of developing diabetes than are people from diabetes-free families.

INSTRUCTIONAL GOAL: BREAK DOWN AND DEFINE COMMON MEDICAL TERMS USED FOR SYMPTOMS, DISEASES, DISORDERS, PROCEDURES, TREATMENTS, AND DEVICES ASSOCIATED WITH THE ENDOCRINE SYSTEM.

Signs and Symptoms

Prefix	Definition	Combining Form	Definition	Suffix	Definition
ex-	*outside, away from*	acid/o	*a solution or substance with a pH lower than 7.0*	-ia	*condition of*
poly-	*excessive, over, or many*	acr/o	*extremity*	-ism	*condition of*
		dips/o	*thirst*	-osis	*condition of*
		ophthalm/o	*eye*	-megaly	*abnormally large*
		ket/o	*ketone*	-s	*more than one*
		hirsut/o	*hairy*	-uria	*urine, urination*

Medical Term	Definition
Acidosis	an abnormal accumulation of waste materials that are acidic, often a symptom of diabetes mellitus; acidosis also can be caused by respiratory or kidney disorder
Acromegaly	an enlargement of bone structure most prominent in the face and hands, resulting in disfigurement, and caused by the hypersecretion of growth hormone from the pituitary gland after puberty
Exophthalmos	abnormal protrusion of the eyes
Goiter	an abnormal enlargement of the thyroid gland caused by a tumor, lack of iodine in the diet, or infection
Hirsutism	excessive body hair in a masculine pattern
Ketosis	excessive amount of ketone bodies in the blood and urine, which is a symptom of an abnormal metabolism of carbohydrates as seen in uncontrolled diabetes and starvation; also known as ketoacidosis
Polydipsia	an abnormal state of excessive thirst
Polyuria	the excretion of abnormally large volumes of urine

Diseases and Disorders

Prefix	Definition	Combining Form	Definition	Suffix	Definition
endo-	*within*	aden/o	*gland*	-al	*pertaining to*

hyper-	*excessive, abnormally high, or above*	adren/o	*adrenal gland*	-emia *condition of blood*
hypo-	*deficient, abnormally low, or below*	calc/i, calc/o	*calcium*	-ism *condition of*
para-	*alongside or abnormal*	carcin/o	*cancer*	-it is *inflammation*
		crin/o	*to secrete*	-oma *tumor*
		glyc/o	*sweet or sugar*	-pathy *disease*
		gonad/o	*sex gland*	
		myc/o	*mucus*	
		pancreat/o	*pancreas*	
		thyr/o, thyroid/o	*shield, thyroid*	

Medical Term   **Definition**

Addison's disease   a chronic syndrome caused by hyposecretion of the adrenal cortex,

characterized by darkening of the skin, loss of appetite, mental

depression, and muscle weakness

Adenitis	inflammation of a gland
Adenosis	abnormal condition of a gland
Adrenalitis	inflammation of the adrenal gland
Adrenomegaly	abnormal enlargement of the adrenal glands
Adrenopathy	general disease of the adrenal gland
Cushing's syndrome	a syndrome resulting from hypersecretion of the adrenal cortex, characterized by obesity, moon face, hyperglycemia, and muscle weakness
Diabetes insipidus	Caused by hyposecretion of ADH by the posterior lobe of the pituitary gland, its symptoms include polydipsia and polyuria.
Diabetes mellitus	A chronic disorder of carbohydrate metabolism, it includes type I, which requires hormone replacement therapy with insulin, and type II, which usually can be managed with diet and exercise program.
Endocrinopathy	general disease of the endocrine system
Hyperadrenalism	excessive activity of one or more adrenal glands
Hypercalcemia	abnormally high calcium levels in the blood
Hyperglycemia	abnormally high glucose levels in the blood
Hyperinsulinism	excessive amounts of insulin in the blood, which pulls sugar from the blood, resulting in low levels of sugar in the blood, fainting, and convulsions

Hyperkalemia	abnormally high potassium levels in the blood
Hyperparathyroidism	hypersecretion of the parathyroid glands, usually due to a tumor
Hyperthyroidism	hypersecretion of the thyroid gland, characterized by exophthalmos, goiter, rapid heart rate, and weight loss; also called Graves' disease, or thyrotoxicosis
Hypoadrenalism	abnormally low adrenal activtity
Hypocalcemia	abnormally low calcium levels in the blood
Hypoglycemia	abnormally low blood sugar levels
Hypokalemia	abnormally low potassium levels in the blood
Hyponatremia	abnormally low sodium levels in the blood
Hypoparathyroidism	hyposecretion of the parathyroid gland
Hypothyroidism	hyposecretion of the thyroid gland, characterized by slow heart rate, dry skin, low energy, and weight gain
Myxedema	advanced hypothyroidism in adults, with the characteristics of low energy, swollen hands and face, and dry skin
Pancreatitis	inflammation of the pancreas
Parathyroidoma	tumor of the parathyroid gland
Pituitary dwarfism	caused by hyposecretion of growth hormone by the pituitary gland at an early age, slowing growth and causing a short but proportional stature; it is a congenital condition that can be treated during childhood with

growth hormone therapy

Pituitary gigantism	caused by excessive secretion of growth hormone that begins before puberty, leading to abnormally increased growth of bones to produce a very large stature
Syndrome	a disorder with an array of symptoms, involving multiple organs
Thyroiditis	inflammation of the thyroid gland

Treatments, Procedures, and Devices

Prefix	Definition	Combining Form	Definition	Suffix	Definition
endo-	*within*	adren/o	*adrenal gland*	-al	*pertaining to*
para-	*alongside or abnormal*	crin/o	*to secrete*	-ectomy	*surgical excision or removal*
		thyr/o, thyroid/o	*shield, thyroid*	-logy	*study of*
				-oma	*tumor*
				-tomy	*incision or to cut*

Medical Term	Definition
Adrenalectomy	excision of the adrenal gland
Fasting blood sugar	a diagnostic test to determine blood sugar levels following a 12-hour fast; extreme variations in blood sugar are an indication of diabetes mellitus
Endocrinologist	a physician specializing in the treatment of endocrine disorders
Endocrinology	a field of medicine focusing on the study and treatment of endocrine disorders
Endocrinopathy	a disease or disorder in the function of an endocrine gland, and the related consequences
Glucose tolerance test	a diagnostic test to confirm a diagnosis of diabetes mellitus and to determine other abnormalities in glucose metabolism; usually following an FBS, the patient is given glucose either orally or intravenously; then, at timed intervals, food samples are taken, and glucose levels are determined and charted. Postprandial blood sugar is a measurement of blood sugar levels after a meal, usually at two-hour intervals.
Radioactive iodine uptake	a diagnostic procedure measuring thyroid function, in which radioactive iodine uptake into the thyroid gland is measured. This is a diagnostic procedure that records an image of the thyroid gland following the oral administration of a labeled substance,

usually iodine; used to detect thyroid tumors

Thyroxine test	a direct measurement of the amount of thyroxine in the blood to determine hyperthyroidism or hypothyroidism
Hormone replacement therapy	the use of a drug that replaces a hormone to correct a hormone deficiency
Parathyroidectomy	excision of one or more parathyroid glands
Postprandial blood sugar	a measurement of blood sugar levels after a meal, usually over two hours
Radioiodine therapy	the use of radioactive iodine to treat a disease of the thyroid gland, such as a thyroid tumor
Thyroidoma	a thyroid tumor
Thyroidectomy	excision of the thyroid gland
Thyroidotomy	incision into the thyroid gland
Thyroparathyroidectomy	excision of the thyroid and parathyroid glands
Thyroxine test	a dignostic test used to measure thyroxine levels in the body

Factoid

Consistency is very important to diabetics. As patients' activity changes, their insulin requirements change. This is why it is so important for diabetics to test their blood sugar frequently each day, especially if they are physically active.

Factoid

The parathyroid hormone stimulates the following functions: release of calcium by bones into the bloodstream, absorption of food by the intestines, and conservation of calcium by the kidneys. Clinical results suggest that diabetes might be treated by islet transportation early in the clinical course of the disease before the development of complications and without the risk associated with conventional immunosuppression.

Factoid

Hypoglycemia means "low blood sugar." Hyperglycemia means "high blood sugar." Why is it that we do not fluctuate from one to the other between meals? The answer lies in the function of insulin. Hormones influence or antagonize insulin by increasing or decreasing the level of glucose in the bloodstream, helping maintain a constant level. The body is maintained in a state of normal and healthy homeostasis. Good health and life itself rely on balance.

Abbreviations.

Abbreviation	Definition
ADH	antidiuretic hormone
DI	diabetes insipidus
DM	diabetes mellitus
FBS	fasting blood sugar

FSH	follicle-stimulating hormone
GH	growth hormone
GTT	glucose tolerance test
HRT	hormone replacement therapy
LH	luteinizing hormone
PH	parathyroid hormone
PPBS	postprandial blood sugar
RAIU	radioactive iodine uptake

CHAPTER FOURTEEN

The Endocrine System

Worksheet 1

Phonetic Spelling Challenge

Spell the medical term correctly in the space provided.

1. miks eh DEE mah   _____

2. pih TOO ih tair ee DWARF izm   _____

3. THYE royd EYE tiss   _____

4. en doh krin ALL oh jist   _____

5. HIGH poh kal SEE mee ah   _____

6. HIGH poh GOH nad izm   _____

7. EN doh krin AH path ee   _____

8. PAN kree ah TYE tiss   _____

9. DYE ah BET ik nef ROHP ah thee   _____

10. HIGH poh add REN al izm   _____

Spelling Challenge

These terms are spelled incorrectly. Spell each term correctly in the space provided.

1. Pancreaz   _____

2. Thieroid   _____

3. Adison's disease _____

4. Thyroidtoxicosis _____

5. Parofthyroid glands _____

6. Addenoma _____

7. Diabetees incipidus _____

8. Pollydipsia _____

9. Acidoses _____

10. Acromeagally _____

Abbreviation Matchup

Select and match the correct abbreviation to the definition.

_____ **1.** diabetes insipidus   **a.** GH

_____ **2.** follicle-stimulating   **b.** FBS

hormone

_____ **3.** growth hormone   **c.** DM

_____ **4.** fasting blood sugar   **d.** DI

_____ **5.** glucose tolerance test   **e.** FSH

_____ **6.** diabetes mellitus   **f.** GTT

True/False

Mark each statement as true (T) or false (F).

_____ **1.** The failure of an endocrine gland to produce sufficient levels of a hormone, or hyposecretion, can have a serious impact on health.

_____ **2.** The term *endocrine* literally means "to secrete within."

_____ **3.** In the condition known as myxedema, the skin becomes loose and thin.

_____ **4.** "Hyper-" means excessive.

_____ **5.** "Hypo-" means below normal.

_____ **6.** Pancreatitis is an acute reaction to infection or trauma and can become life-threatening.

_____ **7.** In general, any disease affecting the endocrine system is called an endocrinopathy.

_____ **8.** HRT may be used following the surgical removal of an endocrine gland to restore homeostasis.

_____ **9.** Thyroxine regulates glucose metabolism and cell division in most cells of the body.

_____ **10.** A reduction of iodine uptake is an indication of deficient thyroid function.

Fill in the Blank: Fill in the blank with the correct medical term from this chapter.

11. A test that may be used to confirm a diagnosis of diabetes mellitus and that examines a patient's tolerance of glucose is called _____.

12. The protrusion of the eyes is a common symptom of _____.

13. The surgical removal of the parathyroid glands with the thyroid gland is called _____.

14. The surgical removal, or excision, of a parathyroid gland is called a(n) _____.

15. A symptom of excessive body hair is known as _____.

16. The abnormal protrusion of the eyes is known as _____.

17. Excessive production of androgens in women also can lead to muscle and bone growth. The resulting pattern of masculinization is known as _____.

18. The prefix "poly-" means _____.

19. A sign that includes enlargement of bone structure is known as _____.

20. A child suffering from the thyroid gland's inability to produce normal levels of growth hormone at birth can develop the condition called _____.

Short Answer

Write the definition for each of the following terms.

21. Adenocarcinoma _____

_____

22. Thyroiditis _____

_____

23. Thyroid scan _____

_____

24. Hypercalcemia _____

_____

25. Myxedema _____

www.ingramcontent.com/pod-product-compliance
Lightning Source LLC
Chambersburg PA
CBHW080007210526
45170CB00015B/1873